Knowledge in the Development of Economies

NEW PERSPECTIVES ON THE MODERN CORPORATION

Series Editor: Jonathan Michie, *Director, Department for Continuing Education and President, Kellogg College, University of Oxford, UK*

The modern corporation has far reaching influence on our lives in an increasingly globalised economy. This series will provide an invaluable forum for the publication of high quality works of scholarship covering the areas of:

- corporate governance and corporate responsibility, including environmental sustainability
- human resource management and other management practices, and the relationship of these to organisational outcomes and corporate performance
- industrial economics, organisational behaviour, innovation and competitiveness
- outsourcing, offshoring, joint ventures and strategic alliances
- different ownership forms, including social enterprise and employee ownership
- intellectual property and the learning economy, including knowledge
- transfer and information exchange.

Titles in the series include:

Corporate Governance, Organization and the Firm
Co-operation and Outsourcing in the Global Economy
Edited by Mario Morroni

The Modern Firm, Corporate Governance and Investment
Edited by Per-Olof Bjuggren and Dennis C. Mueller

The Growth of Firms
A Survey of Theories and Empirical Evidence
Alex Coad

Knowledge in the Development Economies
Institutional Choices Under Globalisation
Edited by Silvia Sacchetti and Roger Sugden

Knowledge in the Development of Economies

Institutional Choices Under Globalisation

Edited by

Silvia Sacchetti

Institute for Economic Development Policy, The Birmingham Business School, University of Birmingham, UK

Roger Sugden

Stirling Management School, University of Stirling, UK

NEW PERSPECTIVES ON THE MODERN CORPORATION

Edward Elgar

Cheltenham, UK • Northampton, MA, USA

Published by
Edward Elgar Publishing Limited
The Lypiatts
15 Lansdown Road
Cheltenham
Glos GL50 2JA
UK

Edward Elgar Publishing, Inc.
William Pratt House
9 Dewey Court
Northampton
Massachusetts 01060
USA

A catalogue record for this book is available from the British Library

Library of Congress Control Number: 2009930874

FSC
Mixed Sources
Product group from well-managed forests and other controlled sources
www.fsc.org Cert no. SA-COC-1565
© 1996 Forest Stewardship Council

ISBN 978 1 84844 115 6

Printed and bound by MPG Books Group, UK

Contents

Figures

Tables

Contributors

Giacomo Degli Antoni, EconomEtica, interuniversity centre of research, Bicocca University, Milan, Italy

J. Robert Branston, School of Management, University of Bath, UK

Philip Cooke, Centre for Advanced Studies, Cardiff University, UK and School of Development Studies, University of Aalborg, Denmark

Sonja Grönblom, Department of Economics, Åbo Akademi University, Åbo, Finland

Alberto Ianes, European Research Institute for Cooperative and Social Enterprises (EuRICSE), University of Trento, Italy

Francesco Sacchetti, Department of Political Science and Sociology, University of Florence, Italy

Silvia Sacchetti, Institute for Economic Development Policy, The Birmingham Business School, University of Birmingham, UK

Lorenzo Sacconi, Department of Economics, University of Trento and EconomEtica, interuniversity centre of research, Bicocca University, Milan, Italy

Kenneth C. Shadlen, Development Studies Institute (DESTIN), London School of Economics and Political Science, UK

Roger Sugden, Stirling Management School, University of Stirling, UK

Robbin Te Velde, Dialogic Innovation and Interaction, Utrecht, Netherlands

Ermanno Tortia, Department of Economics, University of Trento and European Research Institute for Cooperative and Social Enterprises (EuRICSE), Italy

Johan Willner, Department of Economics, Åbo Akademi University, Åbo, Finland

James R. Wilson, Orkestra-Basque Institute of Competiveness, and University of Deusto, Spain

Preface

This book is part of an ongoing long-term project, which began in our minds about early 2005. It started as the exploration of an intuition about issues that, at that time, seemed to us to be relevant for understanding economic systems. Since then and the time of writing (early 2009), a number of significant changes have happened across economies and in our own, practical histories, and those changes have contributed to the shaping of the volume. Indeed, in certain respects we did not dare to disconnect the reflection about humans and their choices from an assessment of the significance of people's practical experience, from the way in which – to quote Bertolucci – humans move through time in the practical and in the historical sense.

The four years since the inception of the volume have enriched its final form with the perspectives of those who have experienced personal reflection and debate, at times in quite unusual contexts, about a common area of interest. As academics we have chosen to explore knowledge-related issues stemming from our background and from a specific concern for the governance of production activities and the development of localities. And with localities and the people within, we have started to interact in an original – or perhaps primeval – way: us looking at them and them looking at us, searching for what academia and other 'publics' had in common when conceiving choices about development.

In particular, this was possible thanks to the First Biannual International Festival on Creativity and Economic Development, which was held in Gambettola (Emilia-Romagna, Italy) in May 2007, and whose second edition is on the way as we write. Our appreciation and recognition goes to the Mayor of the town, Iader Garavina, and to the entire administration for having embraced the big challenges that such a project implied. Thanks to all the partners who supported the initiative, in particular Raffaella Bassi Neri, Pier Luigi Alessandri, Gabriele Galassi, Angelo Grassi, the Province of Forlì-Cesena and the Legislative Assembly of the Emilia-Romagna Region. Our appreciation goes to the community of the town, and especially to the teachers, the students and pupils of Gambettola's schools for having engaged in the number of activities that accompanied the Festival. Especial thanks to participants in the Festival's Ideas Laboratory, for having experienced academic debate in such an eclectic context; part of the contributions included in this volume stem from that experience.

Thanks also to the participants in the sessions on Economic Analysis and the Role of the Academic convened at the EUNIP Conference 2008 (San Sebastian, Spain), including Johan Willner and James Wilson, both of whom also participated in and supported the Gambettola Festival. Moreover, thanks to the participants in the Dare Workshop 2008 (Åbo Akademi University, Turku, Finland), organised by Sonja Grönblom and Johan Willner.

We are also grateful to Ermanno Tortia, for having read and commented on various chapters of this book; to all of the authors in this volume; to the editor of this special series, Jonathan Michie, for having invited us to submit this project; to the publisher, for having put up with our constant breaking of submission deadlines.

Finally, we would most especially like to acknowledge three people. Marcela Valania, for her role and inspiration in the journey that the three of us have taken together these last few years. Throughout that time we have passionately and intensively discussed issues and promoted activities. Last but certainly not least, Giordano and Alessandro Tortia, who have taught us so much, not least that curiosity and imagination are at the heart of critical enquiry.

Silvia Sacchetti and Roger Sugden
Birmingham and Stirling, 21 January 2009

PART I

Introduction

1. Introduction

Silvia Sacchetti

For the purpose of economic science the process of cumulative change that is to be accounted for is the sequence of change in the methods of doing things, – the methods of dealing with the material means of life.

Veblen ([1898] 1998, p. 410)

1. UNEVEN DEVELOPMENT, COMPETITIVENESS-RELATED 'HABITS OF THOUGHT' AND ECONOMIC INSTITUTIONS

If we had to frame the context of this book, we could evoke concerns about economies, the loss of critical capacity within them and, more broadly with respect to society. The decay of critical thinking in society is, in a sense, the decline of democracy. What we mean is that if we do not exercise our critical thinking and we renounce imagining and possibly pursuing different ideas, different worlds, we lose our interest in participating in a process where objectives are already given. With these issues in mind, this book aims at providing a specific perspective on the role of knowledge for the analysis of economic development issues. In particular, we emphasise knowledge, creativity and critical thinking as forces in processes of change, including those transforming the rules governing economic systems. A treatment of the scope for change, in those terms, is rooted in a concern for economic governance and the enduring imbalances related to the economic system and its institutions.

An earlier analysis of uneven development within the global economic system was provided by Hymer (1972), pointing to the world economy dominated by transnational corporations as a system based on inequality of income, status and authority (see also Pitelis, 2002). The implications of Hymer's analysis have maintained remarkable relevance over the last decades. The majority of the world's economic activities are increasingly coordinated by transnational corporations (UNCTAD, 2008), whose governance is based on hierarchy. Their choices reflect the motivations and objectives of specific and restricted groups which, nonetheless, are extremely influential on the evolution and development choices of localities across the world (Cowling and Sugden, 1998a). Transnationals are actors within a specific economic system, they are disciplined by its rules and beliefs and, yet, they have enough power to influence them. The 'violation of equality and opportunity' in the competitive game (Knight, 1923) reproduces itself not only because of the relative strength of these organisations, but also because of an institutional system which selects the values and behavioural rules that support and accept such imbalances.

The 'habits of thought', as Veblen ([1904] 1996) would say, which prevail in economic institutions (e.g. the business enterprise within the capitalist system) mould the point of view from which other human activities are understood and therefore shaped. Sugden (2004), for instance, emphasises problems related to strategic failure in universities, a specific circumstance where the exclusive definition of research programmes on the basis of, for example, firms' objectives, would deprive academic communities of their independence, subjugating critical analysis to externally defined goals and strategies. This could cause, as a consequence, a divergence between a society's need for advancements in science, and firms' needs to respond to short-term profit rationales.

The selective criterion in such approach is arguably profit-making through competition. In particular, competitiveness and its current understanding is the mechanism of market selection, the one which can mould habits or thought as well as patterns of behaviour and relationships across regions and nations.[1] Within what has been defined as the 'free market system' (see Cowling and Sugden, 1998b for a criticism) dominated by the 'Washington Consensus' (Stiglitz, 2002; Sugden and Wilson, 2002) change to reflect and recognise differences in people's motivations and values might find opposition and be slowed down by previous historical facts which shaped the economic life of regions and localities, including the conventionalities and capabilities which are reflected in and promoted by economic institutions (Veblen, 1898; Nelson, 1994), and which often mirror conflicting interests (Hirschman, 1970).

What, then, are the possibilities, in analytical terms, for turning away from uneven development? Has the Veblenian 'idle curiosity' enough space to possibly change the habits of thought and the rules governing the economic system? Dominated by transnational corporations and framed by Washington Consensus institutions, at present globalisation promotes performance criteria related to competitiveness based on productivity and hierarchical organisations. The market, nonetheless, includes a set of institutions which are potentially capable of hosting different entrepreneurial models, as the case of social cooperatives shows (see Ianes and Tortia in this volume, Chapter 8). For similar reasons, it has been argued that the capitalist economic system should not be confused with the market system (Zamagni, 2005), although the two are often strictly associated because of the prevailing form of organisation acting on the market, i.e. the for-profit enterprise (Borzaga and Tortia, 2009). The market, therefore, although dominated by specific habits of thought, can give space to a plurality of motivations, objectives, behaviours.

However, current competitiveness-related habits of thought design patterns that reinforce a specific business culture, the same which has contributed to raise concerns around the unevenness of development. The question, radically, seems therefore to be how selection criteria can change over time, or why are some patterns preferred to others and how the balance across preferred patterns changes. While we agree with Hodgson (2003) when saying that Darwinism is a precondition of Lamarkian processes of change, not least because even human will is a biological function, we include in our analysis a focus on economic choices that involve individuals and their will (Penrose, 1952; Metcalfe, 1998). Strongly interacting with existing institutions and habitual way of thinking through cumulative causation (Veblen, [1898] 1998), human preferences and choices play a role in shaping economic change. As human will is an emerging property at the individual level, in the same fashion the purposeful creation of institutions can be an emergent property of the social realm. These emerging properties need to be compatible with lower ontological levels (the individual as well as his/her biology). Consistently:

> We argue for the role of knowledge, learning and critical thinking in processes that guide change; in particular, we refer to change of rules and behavioural patterns as well as of underlying habits affecting in a circular flow both institutions and individuals acting within and across economies.[2]

In line with that, in the concluding chapter of the volume we address the complementarities between the private and the public spheres of deliberation, arguing for a need to focus on the 'percorso' that individuals undertake both at the private and at the public level in order to discover their values, habits of thought, inner desires, objectives and preferences. We put forward a pragmatic notion of rationality based on humans' eagerness for understanding and discovering by means of experience (a synthesis between thought and practical action) as an attempt to recapture both the unrestrainable evolution of human wants as well as the space of human interaction, which includes communication and deliberation.

Overall, chapters position knowledge in a number of debates. Part II addresses aspects of knowledge in the organisation of production, analysing the implications for the development of regions and localities. On this basis, Part III deals with institutional and organisational forms that may favour participation of actors into knowledge creation, learning and strategy making. In Part IV, along the same line, knowledge creation and use are reconsidered by focusing upon the role of creativity and creative activities for the enhancement of knowledge, awareness and participation across people and economies.

2. KNOWLEDGE, PRODUCTION AND REGIONAL DEVELOPMENT

Starting the second part of this volume, in Chapter 2, Silvia Sacchetti addresses issues of uneven development by looking at differentials of skills and decision-making power across countries, as a consequence of the division of labour operated internationally by transnational corporations. The chapter offers a critical discussion of the role of knowledge in production activities. The author relates classical economists' ideas of division of labour and technological innovation with more recent perspectives on the theory of transnationals. She looks at the extreme situation in which the division of labour across countries operated by transnational corporations divides regions according to specialisations and, therefore, reflects the same hierarchy of skills that is required to perform specific functions. Such a process potentially perpetrates a 'cumulative causation' process that reinforces uneven development.

Along a similar line, Shadlen (Chapter 3) argues that indeed developing countries' industrialisation is primarily based over learning and adaptation – often subsumed with the idea of technology transfer. Complementary, Sacchetti's discussion of the nature of knowledge and of the different abilities and skills involved with each production activity provides some insights for an understanding of why technology transfer to developing countries – often advocated as a panacea for industrial development – may not only be conditional on a restrictive regulation on the use of intellectual property rights, but may, as well, set up a wider institutional boundary in terms, for example, of government regulation, tax, education and innovation systems. It is argued that such a dynamic may plough an evolutionary trajectory defined by exclusive interests.

More specifically, in Chapter 3, Kenneth C. Shadlen analyses inequalities generated by the intellectual property rights system. He goes to the heart of those international institutions – GATT initially, WTO nowadays – that have progressively established common rules to enable harmonisation of intellectual property (IP) across economic systems. What happens when knowledge or, more precisely, codifiable knowledge becomes privately owned? What the author contends is that recent trends in IP protection along the lines suggested by industrialised countries (i.e. the OECD, the EU, the US) did not contribute to diminish the impact of existing imbalances across countries but, rather, reinforced the unevenness of development. IP harmonisation by regulatory institutions is argued to impose restrictions on developing countries, specifically – for instance – with respect to industrialisation and health care. This outcome, the author contends, is further reinforced by regional and bilateral agreements (RBTAs), which tend to impose even more restrictive regulation than TRIPS (trade-related intellectual property rights). Facilitation of knowledge generation through TRIPS and RBTAs is currently pursued at a global level, and does not account for specific outcomes that endorse regional imbalances. Besides, the author argues that policies towards IP do not recognise the differences in the modality of knowledge acquisition existing between so-called developed and late-developing countries. In particular, Shadlen maintains that the latter tend to acquire knowledge through 'learning, imitation and adaptation', which are precisely the processes that are blocked by current trends in IP regulation.

Philip Cooke (Chapter 4) takes a wider view on the knowledge economy which encompasses potentially all sectors – in contrast with the OECD's notion, which focuses on science-based sectors (OECD, 1996). His suggested approach to uneven development lies in the presence, across regions, of clusters that differ in their ability to innovate by way of mutation and to absorb from the variety of competences emanating from related industries and clusters. In doing so, Cooke does not set aside the possibility for peripheries to develop knowledge-economy activities, provided that marginal regions can construct a network of relationships between so-called 'secondary cities' (in particular, knowledge-intensive business services) and 'primary cities', where high-tech manufacturing is more likely to be developed. The inability to network with high-tech manufacturing is identified as one of the possible determinants of Myrdal's cumulative causation problem (Myrdal, 1957). The specific case-studies on green innovation clusters (North Carolina, US; North Jutland, DK; Wales, UK) introduce, in this chapter, the idea of a transversal innovation superstructure that can serve the process of technological convergence across diverse industries and regions, contributing to clusters' mutation through sectors' cross-fertilisation.

A relevant role in regional and national innovation systems is played by higher education, although literature on the topic has been far more interested in looking at the outcomes of collaborations amongst universities, the public administration and firms, whilst disregarding the rules – implicit or explicit – that regulate academic output. James R. Wilson (Chapter 5) addresses recent trends in the 'commercialisation' of universities – a short-term policy approach to regional and national innovation systems – and analyses its potential impact on development. The latter is understood from a specific angle that puts at the centre the dangers associated with the exclusion of local communities from the definition of strategic choices and argues for the desirability of inclusive decision-making processes (Sugden and Wilson, 2002). Wilson, in particular, discusses universities as organisations that are in a position to feed, promote and develop procedural habits and rules across societies based on critical thought and deliberation, thus enhancing democracy in strategic-choice processes. Current trends towards commercialisation, however, also exacerbate the idea of competitiveness in higher education and threaten their role, whilst favouring governance criteria that mimic those followed by large corporations (Sugden, 2004). The focus on the output rather than on the processes associated with competitiveness might then hold back universities from playing their advocated role, and favour instead exclusion from strategic choice-making. This occurs both within universities – where managerialism is putting academics under exceptional pressures in order to measure performance (see Grönblom and Willner, Chapter 6) – and across society – where universities' role is seen as serving regional competitiveness and productivity in the present rather than stimulating critical and creative capacities across people in the long run.

3. EMERGING INSTITUTIONAL SETTINGS, CRITICAL THINKING AND KNOWLEDGE

Is the evolution of economic institutions going to be determined by the habits of thought and rules defined by market competitiveness, as currently understood? One effect of exclusion is the loss of interest in economic action by most of the people, as Sacchetti and Sugden emphasise in Chapter 9. The third part of this volume discusses alternatives to the patterns and selection mechanism defined by an exclusion-based notion of competitiveness and addresses emerging organisational settings and governance structures that give greater emphasis to people's participation.

Theories of human motivation have further reinforced the limits beyond which the process of exclusion 'cannot be carried without arousing a spirit of rebellion which spoils the game for the leaders themselves, not to mention the effect on the output of products upon which people have become dependent' (Knight, 1923: 605). The effect on productivity of exclusive decision-making is the subject matter of what has been relatively recently called the economics of happiness, which focuses on the impact on people's well-being of, for instance, income, employment and inflation but, mostly, of democratic political institutions (Frey and Stutzer, 2002) or participation and perceived fairness at work (Tortia, 2008).

Sonja Grönblom and Johan Willner, in Chapter 6, focus on institutional choices within universities. They consider the new public management approach and its impact on academics' intrinsic motivation. The authors recognise that motivational pluralism whose denial by the new public management menaces the motivation of academics, and which justifies the need for a real variety of choices across society and its different institutions. The real incentive to boost productivity within universities are to be found, according to the authors, in the degree of interest of activities (namely research, teaching, and administration) as well as in the perceived stability of the job. That view is opposed to topical approaches which focus exclusively on incentives, such as the threat of being punished or the promotion of fierce competition amongst colleagues.

If participation is a result of whether activities light up people's interests and passions, the reverse is arguably true as well: the more individuals can participate and, therefore, infuse their own ideas into the process which defines activities, the greater the motivation to act and achieve good results. Institutional solutions to increase inclusion are currently perfected in for-profit firms as well. Lorenzo Sacconi and Giacomo Degli Antoni (Chapter 7) analyse corporate social responsibility building their analysis on the firm's extended governance in the context of information asymmetries and incomplete contracts. Their approach – unlike Grönblom and Willner – looks at individual motivation and choices within the framework of contractarian theory but, rather than leaving the self-regarding individual in a vacuum, they put forward a specific notion of social capital, defined in terms of dispositions, beliefs and relationships that deepens the understanding of how behavioural rules, trust and reputation emerge in the context of a firm extended governance.

Individual choices and institutional choices interact, supporting and reinforcing each other in the evolution of governance structures. More radically, Alberto Ianes and Ermanno Tortia (Chapter 8) look at the participatory model introduced in the 1970s by Italian social cooperatives to combine a specific awareness of community's interests and entrepreneurial activity. By challenging and totally changing the governance approach of traditional for-profit firms, this chapter offers an alternative view on how the 'business game' can evolve through institutional and organisational settings that may boost people's motivation and the achievement of desired objectives. Specifically, social cooperatives can increase workers' motivations even in the presence of lower wages. This has been achieved in the context of the supply of public benefit services, which characterise non-profit organisations. These experiences contain, however, elements concerning participation which can also shed new light on other organisational contexts.

In Chapter 9, Silvia Sacchetti and Roger Sugden reflect on the role of some specific aspects of knowledge, critical thinking and creativity as part of individuals' characteristics and attitudes. To build the space (physical, institutional, mental) for their use, the authors argue, could stimulate people's interests and, therefore, inclusion in economic choice and action. Whether market selection criteria, as they are currently understood, can change over time is perhaps not just a matter of mutation in individuals' motivations, behaviours and objectives, but rather a matter of recognising variety in people and different publics which, in turn, reflects diversity in motivations, behaviours and objectives. Giving space or providing the institutional solutions to let different approaches and views of the world emerge would, at least, allow a more democratic approach encompassing diversity. Deliberation, in this context, would represent a selection mechanism where creative ideas and different perspectives are critically assessed on the basis of behavioural patterns based – amongst other characteristics discussed in this chapter – on the inclusion of interested publics, on the coherence of arguments, openness and mutual respect. Consistently with a phylogenetic process of change (Hodgson, 1993), under deliberation, knowledge creation and learning take the form of processes that have no predefined outcomes or obligations to favour an established view over what might emerge out of deliberation. Deliberation would free 'idle curiosity' to create new knowledge and stimulate change in the economic system.

Accordingly, creativity, which neo-Schumpeterian evolutionary theory has associated with innovation as the driving force of change in capitalism, assumes here a broader meaning. Creativity can drive change by way of innovation, but can also boost people's motivation once it is recognised as a constituent element of human activities, including economic activity. This notion has clear implications for the way the organisation of production could eventually change, allowing more people to play and enjoy the 'business game' or any other activity. In contrast, if we focus on creativity mediated by innovation, we endorse the view according to which the creativity that matters is the one that passes the market test of profitability. This perspective would perpetuate then the notion of a hierarchy of functions where the highest positions are occupied by those who can – in Schumpeter's terms – exert the entrepreneurial role (Schumpeter, 1912).

If the existing system – as Knight (1923) observed – generates its preferences besides satisfying them, then critical thinking rather than acceptance of system-generated preferences and wants would re-establish a perhaps lost link between individuals' preference formation and patterns of behaviour.[3] Knowledge, in this case, takes also the meaning of awareness. Individuals get to know what they prefer (for instance, through critical thinking and deliberation) and behave and chose accordingly. To do this, however, individuals must have suitable capabilities. Knowing what I prefer, or what I am entitled to by established rights, in other words, is not a sufficient condition to translate my preferences into action (Sen, 1984). Individuals can behave according to their preferences only if they can afford it, that is to say if they have both the capabilities and the freedom to do so. To give the knowledge, institutional space and freedom to people is consistent with pluralism in individual preferences. Moreover, the possibility for individuals to reflect on their preferences and act accordingly would activate a process of 'learning by doing' that would further contribute to the redefinition of individuals' preferences. The education system as well as the media and arts can play, under conditions which favour freedom of enquiry, a crucial role in presenting different, consistent alternatives and scenarios.

Borrowing from the processes based on deliberation and collegiality that originally created academia, processes of knowledge creation and learning across society can also affect change in the economic system. The deliberation mechanism, which contributes to mediating knowledge processes, can fertilise related actors and operate at various levels, within and across organisations, government institutions, and publics. As such, deliberation is not meant to replace the market, but to create space and allow for different perspectives to emerge, including on patterns and selection mechanisms, which do not have to be necessarily identified with the equation productivity/competitiveness.

4. CREATIVE ACTIVITIES: ART, MEDIA, TECHNOLOGY, SCIENCE . . .

The third part of the volume addresses activities which are traditionally recognised as creative, such as media and related technologies, art, as well as academic research. Creative activities are analysed here for their role in stimulating critical thinking and deliberative processes across people. Roger Sugden, Robbin Te Velde and James R. Wilson, in Chapter 10, suggest a focus on communication processes that enhance participation of local communities in the definition of development objectives. In particular, communication is suggested as a foundational element of any processes of learning. The latter takes the particular focus of learning to engage with others in decision-making processes, thus reinforcing a specific development need: learning how to govern democratically. This specific 'know-how' – the authors argue – would impact on the governance of economic systems worldwide, thus changing people's expectations with respect to participation in strategic economic choices. The recaptured capability to exert voice would reflect and be reflected in a parallel freedom and ability in communication media, including visual art and information and communication technologies. Similarly to what has been happening across universities, visual arts have – over the 1980s and 1990s – undergone a process of 'commercialisation', which may hinder their role in stimulating critical thinking. In parallel with what Wilson has argued in Chapter 5 about university's governance, this chapter analyses trends in those organisations and industries that manage the diffusion of art (such as museums), film production and distribution, as well as in communication technologies.

Media are treated specifically in Chapter 11, where J. Robert Branston and James R. Wilson analyse the industry from a strategic decision-making perspective. They contrast their approach with the market-failure perspective, arguing for a focus on media because of its relevance for the economy as a whole. In particular, the authors reason about the uniqueness of such a sector, emphasising its impact on the way people perceive and understand reality and choice. By emphasising the publics and their interests, the authors go beyond the classic industrial policy arguments for privatisation and regulation and suggest, instead, looking at the institutional forms by which the industry is governed, stressing the importance of Hirschman's notion of 'voice' as central for deliberation processes (Hirschman, 1970). The cases of BSkyB and NewsCorp provide an illustration of the issues related to an excessive concentration of strategic decision-making power in the media sector, as well as the impact of exclusive governance on people's understanding and critical thinking.

In contrast with exclusive forms of governance of the likes illustrated by Branston and Wilson about media, the context defined by academia, because of its primary nature, could provide examples of how deliberation can occur through collegiality. A common ground provided by academic knowledge, as well as the ability to critically combine different sources are, across universities, both an input and an output of the deliberation process which is supposed to constantly occur amongst academics. The role of deliberation, in this context, is broader than what a strict focus on economic productivity would entail, because the objective in academia is different: to pursue academic freedom so as to embrace 'the right to search for the truth and to publish and teach what one holds to be true. This right also implies a duty; one must not conceal any part of what one has recognized to be true' (Einstein, 1954, quoted in Child, 2007: 3).

For academia to be deliberative does not mean ceasing to be scientific or that anyone is a scientist. But, as a place where coherent and rigorous thought is allowed, critical thinking capacity, intuition and some degree of imagination (Pascal, 1975 [1660]) should be allowed and even promoted. The reason for this is double. First, only if we recognise differential thinking can we have real discussions (Hodgson, 1993) and encouragement of humans' 'idle curiosity'. Second, people's understanding of new situations is higher if they are given the space to exercise their critical capacity and creativity, as the perspective on research methodologies presented by Francesco Sacchetti (Chapter 12) suggests.

Francesco Sacchetti's chapter addresses creativity in the social sciences, comparing quantitative and qualitative methods. In discussing creativity, he goes back to Chomsky (1964) and his distinction between rule-governed and rule-changing creativity. The idea of performance in research – in contrast to what happens with the managerial approach in universities discussed by Grönblom and Willner in Chapter 6 – is analysed on a substantial basis, revealing the implications of distinct creative processes under different methodological choices. This chapter goes to the heart of the research process, which is what the new public management is trying to govern, paradoxically, through the standardisation of motivations and objectives across researchers: a creative process whose peculiarities should be not only recognised but encouraged to reflect diversity of approaches to the study of reality and the acquisition of novel knowledge.

NOTES

1. A prominent and dominant perspective on competitiveness in a globalised economy is the ability for firms and regions to enter global supply chains. These approaches often fail to take into account the nature of production governance within global supply chains and its implications. From a US perspective, for example, Friedman (2005) depicts the world as a huge supply chain, which functions thanks to a political (the end of communism) and technological (information and telecommunication technologies) convergence since the 1980s. In this 'global, web-enabled playing field' geographical constraints are removed by political and technological diffusion and, therefore, competition between US workers/ firms with the 'rest of the world' is suggested to mirror the selection of the fittest and therefore to be increasingly fierce, if not very conflictual. Such a perspective is rather telling, as it reflects a common approach to competitiveness that has been embraced worldwide by policy-makers when investing in people's training and education.
2. See also Hodgson (1999) on the role of learning as being able to reconstruct the individual's capacities and preferences.
3. See, in particular, Cowling (2006) for an analysis of the impact of corporate advertising on individuals' preferences with respect to the time people spend at work. The author argues that these preferences are induced by corporations' high advertising investments and can be in contrast with people's meta-preferences.

REFERENCES

Borzaga, C. and E.C. Tortia (forthcoming), 'The growing social content of cooperative firms: an evolutionary interpretation', in L. Becchetti and C. Borzaga (eds), *The Economics of Social Responsibility*, London: Routledge.

Child, John (2007), 'Academic freedom – the threat from managerialism', paper presented at the Birmingham Workshops on Academic Freedom and Research/Learning Cultures, University of Birmingham, October 2007.

Chomsky, Noam (1964), *Current Issues in Linguistic Theory*, 6th edn, Paris: Mouton.

Cowling, Keith (2006), 'Prosperity, depression and modern capitalism', *Kyklos*, **59** (3), 369–81.

Cowling, Keith and Roger Sugden (1998a), 'The essence of the modern corporation: markets, strategic decision-making and the theory of the firm', *Manchester School*, **66** (1), 59–86.

Cowling, Keith and Roger Sugden (1998b), 'Strategic trade policy reconsidered: national rivalry vs free trade vs international cooperation', *Kyklos*, **51** (3), 339–58.

Frey, Bruno S. and Alois Stutzer (2002), *Happiness and Economics*, Princeton, NJ: Princeton University Press.

Friedman, Thomas L. (2005), *The World is Flat. A Brief History of the 21st Century*, New York: Ferrar, Straus and Giroux.

Hirschman, Albert O. (1970), *Exit, Voice, and Loyalty*, Cambridge, MA: Harvard University Press.

Hodgson, Geoffrey M. (1993), *Economics and Evolution*, Ann Arbor, MI: University of Michigan Press.

Hodgson, Geoffrey M. (1999), *Economics and Utopia: Why the Learning Economy is Not the End of History*, London: Routledge.

Hodgson, Geoffrey M. (2003), 'The mystery of the routine: the Darwinian destiny of an evolutionary theory of economic change', *Revue Economique*, **54** (2), 355–84.

Hymer, Stephen H. (1972), 'The multinational corporation and the law of uneven development', in Jagdish N. Bhagwaty (ed.), *Economics and World Order from the 1970s to the 1990s*, London: Macmillan, pp. 113–40.

Knight, Frank H. (1923), 'The ethics of competition', *Quarterly Journal of Economics*, **37** (4), 579–624.

Metcalfe, Stanley J. (1998), *Evolutionary Economics and Creative Destruction*, London: Routledge.

Myrdal, G. (1957), *Economic Theory and Underdeveloped Regions*, London, Duckworth.

Nelson, Richard (1994), 'The co-evolution of technology, industrial structure, and supporting institutions', *Industrial and Corporate Change*, **3** (1), 47–63.

Organisation for Economic Co-operation and Development (OECD) (1996), *The Knowledge-Based Economy*, Paris: OECD.

Pascal, Blaise ([1660] 1975), *Pensieri*, XVI edn, translated into Italian by Fausto Montenari, Brescia: Editrice la Scuola.

Penrose, E.T. (1952), 'Biological analogies in the theory of the firm', *American Economic Review*, **42** (52), 804–19.

Pitelis, Christos N. (2002), 'Stephen Hymer: life and the political economy of

multinational corporate capital', *Contributions to Political Economy*, **21** (1), 9–26.
Schumpeter, Joseph A. (1912), *Fundamentals of Economic Development*, Cambridge, MA: Harvard University Press.
Sen, Amartya (1984), *Resources, Values and Development*, Oxford: Basil Blackwell Publishers.
Stiglitz, Joseph E. (2002), *Globalization and Its Discontents*, London: Penguin Books.
Sugden, Roger (2004), 'A small firm approach to the internationalisation of universities: a multinational perspective', *Higher Education Quarterly*, **58** (2/3), 114–35.
Sugden, Roger and James R. Wilson (2002), 'Development in the shadow of the consensus: a strategic decision-making approach', *Contributions to Political Economy*, **21** (1), 111–34.
Tortia, Ermanno C. (2008), 'Worker well-being and perceived fairness: survey-based findings from Italy', *Journal of Socio-Economics*, **37** (5), 2080–94.
United Nations Conference on Trade and Development (UNCTAD) (2008), *World Investment Report 2008: Transnational Corporations and the Infrastructure Challenge*, New York and Geneva: United Nations.
Veblen, Thorstein (1898), 'The instinct of workmanship and the irksomeness of labour', *American Journal of Sociology*, **4** (2), 187–201.
Veblen, Thorstein ([1898] 1998), 'Why is economics not an evolutionary science?', *Quarterly Journal of Economics*, July, 373–97, reprinted in *Cambridge Journal of Economics*, **22**, 403–14.
Veblen, Thorstein ([1904] 1996), *The Theory of the Business Enterprise*, New Brunswick, NJ and London: Transaction Publishers.
Zamagni, Stefano (2005), 'Per una teoria economico-civile dell'impresa cooperativa', in Stefano Zamagni and Enea Mazzoli *Verso una nuova teoria economica della cooperazione*, Bologna, Italy: Il Mulino, pp. 15–56.

PART II

Knowledge, production and regional development

2. The organisation of production and the risk of regional divergence: a perspective on the development of knowledge across economies[1]

Silvia Sacchetti

1. INTRODUCTION

In developed countries the speed of technological change is rapid and the diffusion of new knowledge is facilitated by high educational levels and production competencies. In developing countries a consistent part of the population is illiterate, domestic development of scientific and technical knowledge is absent or proceeds very slowly.

Transnational corporations organise their production across nations and, 'on the surface . . . [they] would seem to be ideal vehicles for helping underdeveloped countries' to improve their relative position with respect to industrialised countries' (Hymer et al., 1966, p. 275). Yet they are playing only a minor role as far as knowledge accumulation in developing countries is concerned. The volume of investments in innovative-related activities performed in developing countries has increased marginally (Patel and Pavitt, 2000). Productivity, which is an indicator of the effects of technological transfer and learning, has not increased consistently across countries and industrial sectors. It has been shown that transnationals stimulate domestic firms to become more productive only in sectors where best practices are consistent with their existing competencies and capabilities (Blömstrom, 1986), 'but that there are no significant transfers of modern technologies' (Blömstrom and Kokko, 2002, p. 13). Moreover as Hymer observed (Hymer, 1970), foreign affiliates, sometimes operate in 'enclaves': inside, modern technologies and products manufactured for foreign markets have not much in common with the backwardness of the external environment. Under these circumstances, spillovers of knowledge and fertilisation of domestic firms through learning do not take place (Kokko, 1994).

The recognition of the link between the international division of labour and knowledge diffusion has been somehow disregarded by the economics of innovation and by development and internationalisation theories. Each of these perspectives has, in turn, addressed a specific aspect of the problem, namely: the nature of knowledge and the conditions under which technological change occurs; the importance of human capital formation and educational policies for promoting development; and the role of knowledge assets for the internationalisation of production either to exploit technological advantages abroad or to benefit from localised knowledge in host countries. In this chapter, these angles are three pieces of the same picture, which are combined together in order to provide a perspective on how the knowledge incorporated in production impacts on the accumulation of localised knowledge.

As mentioned, theory affords knowledge an extremely positive role in the activation of endogenous development processes (UNDP, 1990). We acknowledge this point and we go further. The issue of knowledge formation is not just a matter of how much knowledge is accumulated. It entails also a concern about the quality of knowledge and about who controls the direction that knowledge formation takes across localities. The three theoretical perspectives on knowledge formation and technological change, local development, and the internationalisation of production operated by transnational corporations have not yet been combined to provide an answer to this question.

One of the most significant stimuli to identify differences across localities in terms of their function within the international division of labour comes from Hymer's work (Hymer, 1972). At the heart of his radical perspective is the awareness that the production system dominated by transnationals is defined as a hierarchy of activities and powers associated with those activities. The hierarchical structure of the corporation is mirrored by the 'macrocosm' provided by the international economy so that the application of location theory to the corporate hierarchy of functions suggests concentration of control within the economy.

Building on this perspective, we suggest that the fundamental issue that links internationalisation of production by large corporations and the formation of knowledge across different localities is the uneven distribution of decision-making power with respect to technologies, educational programmes and innovation related activities. This idea is rooted in the strategic decision-making approach which, as Cowling and Sugden (1994) maintain, goes to the heart of how production is carried out. The firm, as a focal point of conscious planning (Coase, 1937), allows for the coordination of production from one centre of strategic decision-making. Consistently, a transnational corporation is a means to control and coordinate production from one centre of strategic decision-making across national borders (Cowling and Sugden, 1994). This suggests the possibility of strategic failure across economic systems, which occurs when the process of strategic decision-making in an economy prevents the achievement of a socially desirable outcome (Sugden, 1997).

In particular, the international coordination of manufacturing functions may imply at various degrees, the 'fragmentation' of production knowledge. The concept of 'specialisation' has a positive meaning and refers to the acquisition and consolidation of technical, organisational and social skills together with the ability to understand, if not undertaking, the production process as a whole. In this sense, as Young ([1990] 1996) argues, specialisation is beneficial both for the locality as a collective entity and for individual actors. With 'fragmentation', by contrast, Young (ibid.) identifies the creation of very narrow functions which require little or no specific technical, organisational, or social knowledge. Each mini-function can be monitored but it will hardly let the general competences and skills of the executers emerge. The link with the overall production process is hardly perceptible (Young [1990] 1996, p. 271).[2] Within the international division of labour operated by transnationals, we recognise the risk of excessive fragmentation, especially in those host countries where transnationals' activities support the emergence of subcontractors of the sweatshop kind. In 'transition countries', the nature of manufacturing activities mainly related to traditional sectors is an example of how foreign capital can be a source of fragmentation that disqualifies labour.

Constraints over individuals' capabilities may not combine with local development objectives, thus leading to strategic failure. With respect to knowledge creation, the implications of strategic failure have been left unexplored. Knowledge formation is a time and capital consuming activity. Developing countries, in particular, have been recognised to have many competing needs while suffering from lack of capital.[3] The creation of a knowledge base is a need which, in the short run, struggles with other priorities. For this reason policy-makers may leave the field open to foreign investors. Where in developing countries the rate of technological change and accumulation of production knowledge is mainly determined by FDI inflows, policy-makers face two issues of strategic importance which both have relevant implications for the direction that development may take within regions. The first is the danger of a domination of production, technological and educational decisions by large corporations. The second relates to the need not just to promote the *rate* at which technological change takes place, but also to be aware of and promote the direction of such a change (Hymer, 1970). This could be done, for instance, by encouraging investments in knowledge creation and diffusion (i.e. through R&D and higher education) that are not only linked to corporate interests but that are meant to increase the choice options of people and to pull regions out of subordination.

Building on these considerations, we would like to deal with some of the limitations and effects of transnational corporations as means to facilitate learning and the diffusion of knowledge in developing countries. We then link the results of this analysis to phenomena of geographical polarisation of knowledge creation activities. Our aim is to provide a framework for understanding the direction of industrial development in the light of the international division of labour operated by transnationals. In particular, we focus on those aspects of production that are related to knowledge, its nature and evolution over time. Our contribution emphasises how production decisions influence the evolution of knowledge assets within firms and territories, and determine (sometimes irreversibly) the evolutionary trajectory of localities. Given the cumulative nature of learning and the close links that relate actors' opportunities with their past experiences, we argue that the technological direction defined by transnationals might not have much to offer to developing countries or, worst, might activate a vicious circle that hamper the capability of developing countries to discover and develop innovations of their own.

In section 2 we sketch some classical economists' perspectives on those aspects of production that are mainly related to knowledge. In section 3 we provide a sectorial characterisation of knowledge creation and innovative related activities. In section 4 we relate FDI to the development of knowledge capital within foreign locations, paying particular attention to developing countries. Then, we apply the theoretical considerations of the previous sections to the development of local production systems, looking at those mechanisms that may favour or hamper processes of knowledge accumulation within localities. Section 5 specifically addresses the problem of strategic failure and polarisation effects in knowledge accumulation and innovation related activities. Section 6 applies Hymer's law of uneven development to the international localisation of activities, and provides an explanation of how the organisation of production discriminates amongst economic systems with respect to knowledge formation. We end the chapter by reconsidering industrial development in the light of the impact of the international division of labour as it is planned by transnational corporations. Policy suggestions follow.

2. KNOWLEDGE IN PRODUCTION: SOME CLASSICAL VIEWS

During the production process, inputs are combined by virtue of five complementary forms of knowledge. One is the knowledge of individuals, which is reflected in their 'skills, dexterity, and judgement'. The second is the knowledge incorporated inside capital goods or, in other words, the technology and the tools used to undertake production activities. The third form of knowledge is organisational, rooted in the routines and practices of the firm.

Production, however, does not occur in a vacuum, and increasing specialisation requires network relations and co-ordination amongst firms. Recent developments in the international organisation of production are characterised by the growing intensity of networks of suppliers and prime contractors which, by virtue of their linkages, extend their knowledge and production potential (Sacchetti and Sugden, 2003). Investments in specific technologies require that firms are committed to a constant interaction with other firms specialised in complementary activities.[4] A fourth form of knowledge is therefore relational, and it directly reflects the capacity of firms to use the knowledge of others by means of co-ordination. The fifth type of knowledge is localised knowledge, or the knowledge diffused within a specific space.

According to Adam Smith, the source of value in production is labour. In particular, the value of a good exchanged in the market is measured by the amount, the degree of hardness and the skills required for its production. The knowledge that a worker must cumulate in order to produce implies his or her involvement in a process of learning. A focus on labour, therefore, stresses the importance of human capital and continuous learning not only for those activities that are directly linked to research and development, but also (Adam Smith would probably say 'especially') for those workers who are directly involved in productive activities. Besides physical capital, investments would also be directed to renew and increase labour's knowledge.

When workers undertake production functions they make use of tools and machineries. These means of production embody the knowledge of those who designed them. In this sense, we understand capital goods as a combination of knowledge and matter (Baetjer, 2000). The boy described by Smith who 'was constantly employed to open and shut alternately the communication between the boiler and the cylinder . . . observed that by tying a string from the handle to the valve which opened this communication to another part of the machine, the valve would open and shut without his assistance, and leave him at liberty to divert himself with his playfellows' (Smith, [1776] 1995, p. 14). As Baetjer (2000) notes, the innovation introduced by the young boy was then installed as a standard technology in subsequent engines, and production activities could benefit from the knowledge that that young worker was able to imprint in the earlier machine.

If Smith emphasised the role of knowledge in production in terms of the division of labour and of individual skills, subsequent contributions focused more directly on the role of machines. So the question of why a firm should have introduced new machineries found an answer with Ricardo in the possibility of the capitalist gaining extra profits (Ricardo, 1819). Machines substitute labour if the cost of the innovation is lower than the cost of the labour force substituted by the new technology. The overall employment per unit of output would diminish whenever the production of machines requires less labour than that which is surrogated by the machine itself. Ricardo's argument was reconsidered by Marx ([1867] 1933) and nourished the still on-going debate about the implications of technological change for employment (see Michie et al., 2002).

Marx, however, went further. The technique (tools) applied to machines substitutes individuals, whose sphere of action shifts from that of physical production to the organisation, co-ordination and control of machineries. In one respect, this impoverishes the knowledge and abilities of individuals who are confined by an extreme fragmentation of work (Marglin, 1974). On the other hand, according to Kern and Schumann ([1984] 1991), the introduction of machines, over time, qualifies labour by substituting operational functions with more qualified tasks. Machines can produce physical goods without the mediation of individuals and individuals' production activities become those related to the design, organisation and control of machines. However, shifts from ordinary labour to forms of more qualified work do not occur homogeneously in the economy. Compensation mechanisms may not be thought of as instantaneous and automatic (Zanetti, 2001). Furthermore, there are a number of tasks that are undertaken with the direct involvement of the labour force in manufacturing activities whose location impacts on countries' trade balances (Panic and Joyce, 1980).

Considering the arguments presented above, classic economists rightly anticipated that the organisation of production impacts on the qualification of labour: tasks that are too fragmented disqualify labour. The value of production is generated by the skills and competence of labour which come about through specialisation (as opposed to fragmentation). In parallel, technological progress is associated with specialisation and with the quality of human capital. Vice versa, technologies, which incorporate knowledge, represent a source of learning. These considerations suggest that the technological characteristics of industrial sectors imply different modalities of organisation of production and different learning possibilities at the individual level, at the level of firms and within localities.

3. THE SECTORIAL CHARACTERISATIONS OF KNOWLEDGE CREATION AND INNOVATION

The intensity of the knowledge involved in different production activities varies according to scientific and technical complexity as well as to the tacit or explicit nature of knowledge. In these respects, we discriminate amongst activities both within the same sector as well as amongst different sectors. Accordingly, differences in the international division of labour can be further appreciated by distinguishing production according to the knowledge intensity of labour, technologies and products.

Knowledge and innovation are not developed across the whole economy indistinctively. As Kondrat'ev ([1926] 1935) and Schumpeter ([1912] 1934) emphasised, technological progress occurs following cyclical waves, during which new technological trajectories emerge. At the same time, there may be sectors going through an expansive phase and, sectors undergoing decline. The hypothesis that sectors differ in the rate and modalities of innovation has been interpreted by Pavitt (1984), who distinguishes industries where firms rely on the technology produced by other sectors from those where firms support production with internal R&D. According to this main criterion, sectors are classified as 'supplier-dominated', 'scale-intensive', 'specialised suppliers' and 'science-based'.

In particular, supplier-dominated sectors are essentially traditional labour intensive sectors (e.g. textiles and clothing, leather and footwear, wood and furniture) where innovations are largely related to processes. Technological opportunities are those determined by the new technologies and intermediate inputs produced by other sectors. The process of innovation is therefore characterised by the diffusion of best practices and innovative intermediate inputs. Knowledge is important to the extent that it allows producers to efficiently introduce the technologies produced elsewhere.

Scale-intensive sectors involve complex manufacturing systems for which both product and process innovations are important. Economies of scale can be obtained both for production and R&D activities. Firms are generally large and are likely to vertically integrate the design/manufacturing of their production technologies. Scale-intensive sectors include transport, the production of electrical durable goods, chemicals, glass and other building material.

Specialised suppliers produce capital goods and focus mainly on product innovation. Firms are generally small and act in strict connection with their clients. These goods are capital inputs for other sectors and incorporate very high technical skills of designers. The knowledge included in products is partly tacit and cumulates over time.

Science-based products complete the taxonomy. These include firms whose production is strictly linked with scientific knowledge (e.g. pharmaceuticals, aerospace, electronics and computer industries). Innovation occurs with the emergence of new technological paradigms. Appropriability of knowledge through patents is high and the innovator often benefits from a temporal advantage which allows him/her to exploit market leadership. Innovation-related activities are formally developed inside R&D centres. These goods are often incorporated as intermediate products by other sectors. Firms are often of large size, although small Schumpeterian innovative firms may constitute an exception.[5]

This interpretation of manufacturing sectors introduces Dosi's hypothesis of technological paradigms and trajectories (Dosi, 1991). According to his contribution, the variety of forms of production organisation can be explained in terms of differences of evolutionary processes amongst sectors. Advances in scientific knowledge open up new *technological opportunities*, which may be translated into innovations. New opportunities, in particular, are defined within technological paradigms (ibid., p. 137). It follows that change is the outcome of a cumulative process of knowledge creation, learning and adaptation which cannot be simply traced back to a reaction to changes in market conditions. A change in relative factor prices such as an increase in the cost of labour, or of any raw material, may not be a sufficient element for introducing labour/raw materials saving technologies. The same can be said for demand fluctuations. This perspective suggests that the context in which technological change occurs is not necessarily linked to the market, but can be understood also in terms of what opportunities are offered by the knowledge already existing.

Whilst some knowledge may be generally available, for instance through the market for technologies, tacit and subjective forms of knowledge, which are built in firms' routines and experience, may largely differ across economic actors. Past experience is important to the extent that new opportunities emerge depending on the contextual knowledge that was cumulated earlier within the firm (scientific, technical and organisational knowledge in its tacit and explicit forms) and on the more general advancements of science and technology, eventually stimulated by changes in relative factor prices. On the one hand, there is a private sphere of knowledge evolution which involves individuals and their subjective sphere, as well as organisations. On the other hand, individuals and firms' learning effects are in turn influenced by the evolution of technological paradigms and general science. These knowledge elements are shared by actors operating within a particular technical or local community. Similarly, the larger set of institutions that supports the established technology and industry is oriented towards sectorial specialisation and can be considered as a further element that inhibits the shift from old to new practices and knowledge. As Nelson (1998, p. 330) reminds us, in the early work of Veblen (1915) as well as in the contributions of Perez (1983) and Freeman (1991) emerges the idea that institutions suitable for a specific set of technologies could be inappropriate for the new. In particular, both Perez and Freeman suggest that, after 1970, the rise of information technologies drove a change towards an institutional setting with respect to those needed before. Industrial sectors and firms are heterogeneous with respect to knowledge assets and that new knowledge, in the form of technologies, routines, human capital, and institutions is path dependent: radical shifts require radical changes in the set of specific assets of firms and institutions. These imply costly reform processes and time.

4. TRANSNATIONALS AND TECHNOLOGY TRANSFER

The impact of production activities on local knowledge is not neutral. The taxonomy theorised by Pavitt, which finds several confirmations in applied analysis, links each sectorial group to specific innovation capacity and size characteristics. In the international economic system, the production of knowledge is increasingly linked to large capital investments that leave space for economies of scale to be realised, especially in R&D. Scale-intensive and science-based sectors are two examples of this. More generally, the size of firms is often significantly correlated with learning and R&D.[6]

Transnational corporations (TNCs) are firms of undoubtedly large size which have strongly impacted on the internationalisation of production. Slaughter notices that from 1979 to 1999

> The ratio of world FDI stock to world gross domestic product rose from 5 to 16 per cent, and the ratio of world FDI inflows to global gross domestic capital formation raised from 2 to 14 per cent. One consequence of this is that an increasing share of developing countries' output is accounted for by foreign affiliates of multinational enterprises. (Slaughter, 2002, p. 9)[7]

However, the impact of transnationals on world production is expected to be even higher. With the diffusion of networked organisations, transnationals control production far beyond the legal boundaries defined by property rights, as planning is extended to aligned suppliers and to the cascade of firms that are linked to first-tier suppliers (Cowling and Sugden, 1994; Nolan et al., 2002). A strategic decision-making approach considers the boundaries of the firm as the pattern of structural influences that the firm has on other actors' strategic decisions (Cowling and Sugden, 1998). A TNC's technological strategy, therefore, touches upon a space that is defined by the TNC's power to direct a number of other networked firms.

How do transnationals allocate their knowledge-creating capacity across regions and countries? Outside home countries, as Vernon (1966) anticipated, the nature of technological activities carried out by transnationals seems to be concerned with the adaptation of products and production processes to local market conditions. In particular, process innovation could be determined by differences in the labour market, according to the perspective offered by the Ricardian hypothesis.

More recently, as other contributors have observed (Narula, 2003), the size of R&D investments and the complexity of the knowledge incorporated by technologies has pushed transnationals to look for other factors such as seeking strategic assets created abroad (e.g. host country's technological developments).[8] In this case, through FDI, the firm exploits the knowledge of the host country. For 'asset-augmenting activities' to take place, local knowledge assets must be consistent with the transnational's aim of reinforcing its technological advantage. Therefore, as Narula (ibid., p. 212) observes, reference is to intermediate industrialising economies and industrialised economies, where the national innovation system of those countries supports top-level R&D and education or offers, at a subnational level, the possibility of benefiting from agglomeration economies.

As empirical evidence shows, however, firms producing for the world market 'may keep most of their technology production close to the home base' even in industrialised countries (Patel and Vega, 1999, p. 154) thus supporting (or at least not contradicting) Vernon's initial hypothesis (Vernon, 1966).[9] Reasonably, FDI inflows in developing countries will provide no evidence of the technology sourcing hypothesis. As Patel and Pavitt (2000) have shown, with the exception of pharmaceuticals, the degree of internationalisation of R&D is not positively correlated with the technological intensity of the industrial sector.

Although research activities are mainly concentrated in home countries, transnationals may represent a channel for the diffusion of existing technologies and practices. Current analysis of the relationships between FDI, human capital and knowledge diffusion focuses on three main aspects. One is the upgrading of production processes by local subcontractors through the introduction of production standards. Related to the use of technical standards are voluntary processes of technological transfer and best practices from the foreign affiliate to local partners. In this case the technological paradigm which prevails in the sector determines the technologies and the knowledge that are needed for a firm to be involved in production networks.

Another aspect of interest is provided by knowledge spillovers from transnationals to local economic actors. In particular, on productivity spillovers, authors recognise that FDI creates a potential for spillovers of knowledge to local firms and the labour force (Blomström and Kokko, 2002, p. 10). Spillovers take the nature of 'non-market interdependence' or external economies. In this specific case, external economies are a 'peculiarity of the production function' as the output of a firm (e.g. domestic firms) 'depends not only on the factors of production . . . utilised by this firm but also on the output . . . and factor utilisation of another firm or group of firms' (Scitovsky, 1954, p. 145). Technological externalities occur, for instance, when domestic firms benefit from human capital formation by foreign affiliates, or when domestic firms upgrade their technologies by virtue of their proximity to more advanced technological realities. In both these examples, the transnational aims at exploiting its ownership advantages, such as patents, trademarks or proprietary technologies in the host country. Vice versa, when it is the local knowledge that is attractive, foreign affiliates can benefit from the specific competencies of the local system, hiring skilled personnel (such as in the Irish case), or being close to centres of scientific or technological excellence (as in the case of Silicon Valley).

The examples of spillovers refer to industrialised countries, where results are, however, mixed (Slaughter, 2002). For instance, one of the most prominent results of earlier studies (Vernon and Davidson, 1979) is that technologies are first transferred to high per-capita income countries, with high literacy rates and well developed manufacturing sectors where, however, they do not always increase domestic firms' productivity.[10] For developing countries, on the contrary, there is very little evidence of spillovers (Slaughter, 2002). In particular, spillovers are concentrated in middle-income developing countries while there is no support for the existence of such effects in countries with the lowest per-capita income (Blomström et al., 1994; Blomström and Kokko, 2002).

This variability of results does not allow us to create a straightforward equation between FDI and knowledge spillovers. One reason, as we mentioned earlier, is that spillovers are externalities which are not encouraged by leader firms. On this point, it has been shown that FDI and technological licensing are higher in host countries with a strong property rights system (Lee and Mansfield, 1996; Maskus and Yang, 2000). Following the argument of the non-voluntariness of spillovers, Mansfield and Romeo (1980) conclude that for joint ventures, when spillovers are more likely, US transnationals transfer only older technologies.

A second reason is that spillovers, as well as strategic forms of technological transfer, may occur if there is a basis of knowledge which allows domestic firms to understand new technological opportunities and to apply them. The innovative capability of economic actors is in their ability to integrate the knowledge of others and adapt it to specific needs, thus generating new, non-redundant knowledge (Cusmano, 2000). The transformation of knowledge flows into innovative solutions is related to what Cohen and Levinthal (1989) have defined as 'absorptive capacity', or the ability to recognise relevant external knowledge, assimilate it and apply it to commercial ends. This ability may be a precondition for new knowledge creation and for the introduction of innovation. The ability to absorb and integrate knowledge builds upon the previous learning experiences in general science and in its technological applications. Thus we find a link between present innovation and past knowledge creation: an innovation that is introduced today depends on the path opened by the research activity started previously (Kondrat'ev, [1926] 1935) and by the technologies that are already in use.[11] Deficiencies in communication infrastructures in developing regions and the scarcity of knowledge capital, which is in turn related to low educational levels of the labour force, are major obstacles to the activation of knowledge flows between firms.

One of the major assumptions of approaches based on path dependence is that there is a degree of irreversibility for which previous decisions (for instance in terms of sunk or switching costs, knowledge basis and learning ability) may have irretrievable effects on the future opportunities of actors at the level of individuals, firms and territories. What happens at the local level with respect to production, training, education and basic research institutions, the financial system and market structure is paramount for the process of knowledge creation and technological specialisation.[12] Thus the decision of headquarters to locate activities with relatively low intensities of knowledge in developing countries has, under particular policy conditions, a restrictive effect on the development possibilities of the local industry and of the locality as a whole.

In those economies where the lack of local capital and knowledge leaves space for foreign investments, production decisions about learning and technological direction are the domain of transnationals. For policy-makers, to rely on technological transfer from foreign affiliates to local subcontractors or on the theoretical possibility of spillover effects implies the risk of excluding developing regions from generating autonomous innovative capabilities. With human capital being perhaps poorly educated, lacking technical and managerial skills, these regions will be included in the international organisation of production to provide cost advantages to foreign investors (e.g. in terms of labour) but will be restricted to traditional, labour-intensive activities characterised by a disqualifying fragmentation of tasks. These are sectors, as Pavitt emphasises, that are not autonomous generators of innovations and that do not promote R&D within the sector.

5. STRATEGIC FAILURE AND POLARISATION EFFECTS IN INNOVATION RELATED ACTIVITIES

The assumption we make, building on path dependence, is that learning and technological dynamics can activate a virtuous circle of knowledge generation and accumulation, which can be beneficial for local systems, both because it augments the value-added produced and because it impacts on the ability of local communities to access a wider range of opportunities and shape their development objectives accordingly.[13] Capital accumulation, however, is not exempted from generating conflicts and continuous tensions, either amongst social groups, regions or nations. Phenomena like these could be interpreted as the result of the eventual discrepancies that arise between the objectives motivating firms' strategies and the consequences (more or less unintended) that are generated at the collective level. For example, we can talk about positive unintended consequences in the case of spillovers or about negative (alternatively intentional or unintended) consequences when the development of human capital is hampered by a concentration of traditional, labour-intensive sectors.

Within regions, transnational strategies may generate tensions due to strategic failure. Large transnationals, especially within those industrial sectors which heavily rely on R&D activities, convey great financial resources into innovation programmes. Besides, knowledge creation activities are central in the consolidation of oligopolistic or monopolistic positions and transnationals often operate in oligopolistic or monopolistic markets (Dunning, 1993). Thus, in these sectors, control over strategic decisions with respect to innovative related activities and knowledge assets is retained at the heart of transnationals, which determines their broad corporate direction even (but not necessarily) against the will of other actors involved (Zeitlin, 1974), such as domestic firms, subcontractors, labour, trade unions, consumers, and governments. Examples of strategic failure of this kind can be found in pharmaceuticals where, for instance, the incentive to invest in R&D on the illnesses diffused within industrialised countries may be higher than the inducement to invest in research for finding cures to diseases that are endemic in developing countries, where purchasing power is very low and could not compensate the corporation for R&D expenses.

Strategic failure also occurs when transnationals influence the formation of human capital according to their specific needs. Host countries may experience the transfer of simple or complex technologies depending on the level of a number of elements, such as the quality of their human capital and skills. The nature of these technologies defines the quality of competences that can be learned locally, thus determining what have been named the 'technological opportunities' of local actors (Nelson and Winter, 1982). Given the cumulative nature of learning processes the strategic decision of which pieces of knowledge have to be used internationally within host countries impacts on the direction that localised knowledge will take in the future. The result might not be in line with the development objectives of a country thus incurring in strategic failure. For example, when inward investments promote the demand for highly qualified professionals, scientists or specialised technicians, host countries can be enriched, over time, by diffused scientific, managerial and technical knowledge which can be used as a basis for promoting local entrepreneurship and specialisation. Conversely, a transnational might operate as an 'enclave' and siphon off the best trained individuals from the local system, thus leaving domestic firms with shortages of human capital and knowledge. This effect would negatively impact on the development of domestic entrepreneurship. Another example is that of countries which attract labour-intensive activities and specialise in low-value-added production. Once individuals, firms, and institutions learn to perform and deal with low-value-added activities, they may be locked in and lack the conceptual categories to understand the evolution of knowledge in sectors characterised by higher levels of knowledge complexity.

In parallel, tensions *amongst countries and regions* may occur due to polarisation effects. As the dynamics of development follow different speeds, imbalances amongst local systems can generate, as Myrdal (1957) emphasised, effects of attraction and diffusion with respect, for example, to human and capital resources, trade, or social relations. Developed localities usually exert their power of attraction with respect to the resources of less dynamic centres, whilst diffusion occurs from the strongest locality towards neighbouring systems when the push for expansion is more powerful than the attraction coming from the strongest locality. Each change in either of the two directions (attraction or diffusion) generates a cumulative movement, which will be ascending or descending depending on its causal connection with positive or negative collective effects. Adoption of a long-term perspective led Myrdal (ibid.) to the conclusion that a system does not move towards equilibrium of forces but – through a process of circular and cumulative causation that follows one initial effect – the system tends to incrementally depart from equilibrium. In the long run, complementary effects – and not opposite effects – tend to accelerate changes within the system. As a consequence, by virtue of the process of cumulative causation, the concept of so-called 'free markets'[14] would lead to the creation of regional imbalances, rather than being the mechanism to diffuse development.

The international division of labour as planned by transnationals, influences economic systems at different levels and can generate those initial effects that Myrdal identified as the spark of virtuous or vicious circular cumulative causation. It can be the beginning of a successful process of knowledge accumulation or, vice versa, it can be the foundation of a hardly reversible trend towards the settlement of unqualified labour and activities.

With respect to knowledge formation, polarisation can be observed in phenomena of concentration of innovation-related activities in transnationals' home countries and, more specifically, within some OECD countries, namely Japan and Germany. Given the cumulative nature of learning and innovation, OECD countries, especially the USA and Europe (Germany), are likely, in turn, to attract inward foreign investment that performs innovation-related activities (Patel and Pavitt, 1998).

One reason for this can be found in agglomeration advantages, which provide forms of increasing return to scale in research activities. Geographical concentration of scientific and technological activities has been shown to promote further formation of innovative activities (Arthur, 1990; Saxenian, 1994). The importance of geographical proximity is particularly relevant for the diffusion of tacit forms of knowledge. Tacit, unobservable and complex knowledge, unlike codified knowledge, can be transferred only by means of socialisation, which means that actors are engaged in very frequent exchanges, learning things by doing them together. Although there are exceptions (see, for instance, Sacchetti and Sugden, 2005), this process is argued to be easier when actors are located in the same geographical area. The more the technological regime of an industrial sector requires complex and tacit forms of knowledge, the more it will be characterised by geographical concentration and will determine, therefore, polarisation effects. What is more, concentration of innovation-related activities also depends on the localisation of scientific and technological competencies. The location of universities and governmental R&D centres, for instance, may be relevant in those industries whose activities are linked to basic scientific research, such as aircraft, the production of instruments, motor vehicles and the computer industry.[15] Industrial agglomerations populated by firms undertaking private R&D activities are attractive for sectors where knowledge complementarities require the coordination of research efforts amongst firms or when firms can benefit of pecuniary or technological externalities, such as when part of research findings spill over.[16]

Countervailing forces, in parallel, can initiate a process of knowledge diffusion. Centrifugal effects may arise as a consequence of internationalisation strategies, which may pursue a variety of advantages, namely ownership, location and internalisation (Dunning, 1993).[17] Transnational corporations, in particular, contribute to spread technologies and best practices across their networks of suppliers. However, the nature of the knowledge that is passed on to domestic firms and localities depends on the law of division of labour, with the risk of promoting knowledge fragmentation instead of specialisation.

6. THE RATE AND DIRECTION OF TECHNOLOGICAL CHANGE IN DEVELOPING COUNTRIES

The polarisation of innovation-related activities carries important implications for the location dynamics and structure underlying the international division of labour. Conscious coordination of production by transnational corporations is pushed, on an international scale, beyond firms' boundaries. Planning occurs between firms. In particular, transnational corporations increasingly operate as network organisations, directing and coordinating the activities of a cascade of subcontractors and suppliers. The rate and direction of innovative activities is consistently planned worldwide by a number of large transnational firms, which between the 1980s and mid-1990s increased the proportion of innovative activities performed abroad by only 2.4 per cent (Patel and Pavitt, 2000). Whilst major transnationals have great financial resources and are at the forefront of R&D, developing countries experience problems linked to the lack of financial capital and limited internal market. Inward investments in low-income developing regions can be hardly oriented towards the establishment of innovation-related activities.

Innovation-related activities (such as R&D) are located in some countries whilst they exclude others. However exclusion can be observed also when looking at ordinary production activities. The international division of labour discriminates between countries and territories by allocating production activities that differ by virtue of their knowledge content. Firm's activities differ by the technology used, by the qualification of labour, and by the presence of educational programmes within firms. These characterisations are reflected in the nature of directed technology transfers, and in the nature and magnitude of possible spillovers.

The scenario of international production is compatible with a structure that divides actors into 'superior' and 'subordinate'. As Hymer (1972) emphasised, in a world economy dominated by large transnational firms, the international division of labour is divided into three levels, from the top which is concerned with strategic planning, to the lowest, which is concerned with day-to-day events. This view advances very important welfare implications in terms of 'income, status, authority and consumption patterns'. While skilled workers and superior communication systems are a prerogative of the major centres hosting the first levels of activities, an unskilled labour force characterises those activities related solely to the presence of raw materials, markets and manpower. This means that there are host territories where, although activities are complementary to those of other localities, the level of knowledge involved in production is not high enough to pull actors out of subordination.[18]

Whilst Hymer's uneven development is caused by factors that are external to localities, other contributions emphasise endogenous resources and capabilities as the main determinants of development. As regards individual actors, for instance, differences in the learning capacity imply the existence of organisations where knowledge acquisition or production is poorer.[19] The same principle can be observed also within localities. Systems that are better able to recognise opportunities and learn from experience will gain an advantage with respect to less dynamic and receptive localities.

However, the two causal dimensions (exogenous and endogenous) may be subject to a vicious circle. Localities with poor concentration of knowledge assets and, presumably, decision-making centres, have less strategic decision-making power than localities with superior resources. At the same time, this relative lack of power hinders the possibilities of weak localities being evenly included in the dynamics of knowledge diffusion and creation. If such a circle is activated, the impact of the technological direction planned worldwide by transnationals becomes a very influential element that underlies both the exogenous and endogenous determinants of uneven development.

7. POLICY IMPLICATIONS AND CONCLUSIONS

Production decisions are mainly taken by firms and, at the international level, by transnational corporations. The division of labour across firms and localities, in particular, has been accentuated by the increasing complexity of knowledge contents in production. Complementary activities are compatible, however, with a hierarchy of functions across localities that is characterised by different levels of knowledge contents and by different degrees of economic power.

The evolution of localised knowledge is strongly influenced by the characteristics of its production activities. In particular, local resources and production decisions of firms exert a reciprocal influence on each other. On the one hand, local systems offer specific knowledge assets that may attract production activities. On the other hand, firms also settle their activities on the basis of location advantages. This mutual influence generates a process of cumulative and circular causation between the accumulation of resources and the production functions localised on a territory. Localities that are typified by labour-intensive activities and limited knowledge assets will attract activities that require manpower without highly qualified competences. The resources of a local system – in terms of the knowledge embedded in technologies, learning and research abilities, and relationships – will not expand. In contrast, host countries where specific knowledge has been cumulating over time will attract firms because of their knowledge resources. If the knowledge of foreign affiliates is then spread outside, the high knowledge content of production activities located within the system will further improve the amount of technological, human and relational resources of the territory.

The problem of knowledge creation, diffusion and use in developing countries cannot be solved by totally relying on inward investment flows from transnationals. Although FDI may, within particular institutional contexts, eventually stimulate the demand for qualified labour (e.g. managers, engineers, researchers), there are elements for arguing that within the international division of labour there are activities with different knowledge contents that are associated with specific locations.

First, innovative activities are mainly retained in the country of origin, whilst if the host market is large enough, R&D is performed abroad to adapt product or process to different market conditions.

Second, the innovativeness of the technologies transferred to host countries can be assessed according to two reference points. One is the technological endowment available in the country of origin. The other is the knowledge and technological capital of the host country. With respect to the first, there are no significant transfers of modern technology and domestic firms can at best become more productive in sectors where the technology used is consistent with their capabilities. When comparing foreign technologies with domestic ones, however, there may be a technological upgrading. This suggests that the gap between foreign and domestic technology may be there to stay. Technological transfer to transnationals' suppliers follows the rate and direction decided by headquarters, whilst eventual spillovers can take place only in sectors where domestic capabilities allow local firms to understand and apply the hints leaking from foreign affiliates.

Third, thanks to agglomeration economies, innovative environments attract more investments, thus drawing off resources from other economic systems. This effect can promote knowledge accumulation where the quality of production activities in terms of technologies, R&D and human capital is high. Conversely, localities with poor knowledge assets will not activate a virtuous cumulative process, thus enlarging the initial gap.

Processes of cumulative causation of this sort, as Myrdal (1957) maintained, hamper convergence amongst regions and localities, enlarging the gap between dynamic localities where capital has been consolidated over time and localities where resources have remained poor. Where existing knowledge attracts new resources, virtuous cumulative processes will promote further accumulation, whilst where resources are scarce and knowledge does not spread outside firms, the dynamics of technological change and learning will be jeopardised by firms' strategic choices and by the power of attraction of more advanced areas.

Although these observations are far from being conclusive, they provide some justification for policy which compels developing countries to be aware of the limits of foreign capital with respect to local knowledge development. Transnationals possess undoubtedly great financial resources. However to say this is not 'the same thing as saying that they serve the general interest as well as their own, that they are the best way to exploit the possibilities of modern science' (Hymer, 1970, p. 443). Likewise, when promoting the demand for more qualified labour, transnationals' needs impact on education policies. The risk is that countries dominated by foreign investments 'develop a branch plant outlook, not only with reference to economic matters, but throughout the range of governmental and educational decision making' (ibid.).

Where foreign investment prevails, the rate and direction of technological change will be decided by transnationals to suit their own interests.[20] This contributes to define the nature and direction of localised knowledge and commits technological and human capital development to the opportunities defined by the corporate strategy. This scenario, as the theory of path dependence suggests, may have irreversible effects. Once a locality, together with its institutions and organisations, has committed to a specific learning and research strategy that is functional to the corporate interests of foreign capital, a change in direction would imply a change in the nature of the relationships that have been established between firms and local institutions. Moreover, this would imply the use of existing competences to search for new knowledge. This can enlarge the opportunities of domestic firms and entrepreneurship. However it requires time, capital and, most important, the ability of local actors to effectively define local development goals.

Local institutions may play a major role in relaunching the possibility for developing a knowledge-based production. A fundamental step, not surprisingly, should be towards the enhancement of education. Although in this chapter we have been talking specifically of knowledge across industrial sectors, concerns over the economic emancipation of regions and localities would call for promoting a wider reflection, for instance, on the role of universities but also of different artistic expression within society (on this, see Sacchetti and Sugden, Chapter 9 in this volume). In contrast with approaches that would stress the functionality of universities with respect to corporate objectives (on which see Wilson, Chapter 5 in this volume), we would emphasise, in line with Docherty (2008), the role that universities and higher education can play in developing critical capacity and awareness within and across localities. Such an approach to education and the arts, as argued elsewhere in the volume, would represent a precondition for local development objectives to be endogenously identified. Institutions, far from being irrelevant, may also consistently support this process across industries, for instance by being active in conveying the knowledge produced locally and internationally into production and in creating new specialisations. This could be done by promoting and giving space to criticism and creativity amongst people so that an endogenous debate over development objectives can be activated, as well as through a continuous tension towards innovation and specialisation as opposed to fragmentation.

NOTES

1. This chapter is based on 'Knowledge caps in industrial development' *New Political Economy*, **9**(3), September 2004, pp. 389–412.
2. On the risk of fragmentation see also Marglin (1974).
3. Max Weber ([1905] 2001) emphasised the importance of religious ethics over consumption, whilst Duesenberry (1949) stressed the importance of imitation of externally consolidated consumer habits (the so-called 'demonstration effect'). These contributions gave important instruments for the interpretation of phenomena such as the lack of capital accumulation, under-investment, and to the related obstacles that countries with problems such as those mentioned above have to face to increase their production capacity. See also: Nurkse (1953).
4. On the relationship between investment and co-ordination see Richardson (1960).
5. The description of Pavitt's categories is based on Dosi (1991, p. 147).
6. Of course the role of TNCs in the production of knowledge is not exclusive. Knowledge is produced also by firms that are not transnationals and which may be of small or medium size. The latter aspect is not being addressed in this work.
7. The source of the data quoted is UNCTAD (2000).
8. For a systematisation of the purpose, scale and major determinants of foreign technological activities, see Patel and Vega (1999).
9. For empirical evidence see Love (2003).

10. For two examples of contrasting results, see Chung et al. (2003) and Haskel et al. (2002).
11. Technological changes occur within the trajectories defined by specific paradigms and changes of paradigms occur when innovations are so radical that they break the pre-existing trajectory and subtract economic value from previous technologies; Such radical changes are somehow rare and prediction models inspired by Kondratieff's economic cycles (Kondrat'ev [1926] 1935), have assessed the length of the wave for different industries; On this, see Silvestri (2001). In both cases, innovative processes are based on pre-existing knowledge, either in the forms of paradigms or, in the case of radical innovations, on the previous inventions or research results. On radical innovation, see Schumpeter ([1912] 1934).
12. On the role of institutional actors with respect to the development of countries' knowledge assets, see Lundvall (1992), Nelson (1998), Patel and Vega (1999).
13. Knowledge assumes here another meaning with respect to productive knowledge. It is seen as the perception and recognition of the impact that individual action exerts at the collective level. When this kind of knowledge is institutionalised at the collective level (for instance, through norms and rules which govern economic interactions), local institutions can be better able to identify behavioural patterns that are not consistent with local development objectives and feedback into decision-making processes. On the use of critical analysis and thinking across economies see Sacchetti and Sugden, Chapters 9 and 13 in this volume.
14. Neoclassic economic theory has addressed economic development from a perspective based on the concept of stable equilibrium. Not least, this powerful concept has been at the basis of the theory of so-called 'free markets', where markets have been considered the most effective mechanisms that lead to stable equilibria. In this context the notion of stable equilibrium has been used as an ideal reference point towards which economic systems should be oriented and, as a consequence, equilibrium has been used as a measure to formulate value judgements about the development of economic systems. One problem however seems to remain. Stable equilibrium theories, as argued by Myrdal (1954) have not succeeded in explaining differences between regions and nations in terms of economic development. Critiques on the concept of free markets are based on the idea that power is a constituent element of economies and that economic actors do not have equal power when interacting on the market (see Sacchetti and Sugden, 2003).
15. These sectors have been identified by Patel and Pavitt (1998, p. 304), through an analysis of the geographic location of large firms' US patenting activities.
16. The degree of non-observable knowledge as well as the effectiveness of the legal framework which regulates appropriability of innovation plays an important role in regulating the magnitude of this kind of externality. See Shadlen (Chapter 3 in this volume) for a critical perspective and analysis.
17. However, the strategies pursued through FDI can be interpreted from other perspectives, namely 'divide-and-rule' strategies and imitation and risk reduction strategies. For a deeper treatment, see Pitelis and Sugden (1991).
18. Specialisation as a means to organise production has been explained by neoclassical economics in terms of efficiency. However – expanding on Adam Smith ([1776] 1976) – Marglin (1974) has emphasized that the choice of the division of labour lies 'between the workman whose span of control is wide enough that he sees how each operation fits into the whole and the workman confined to a small number of repetitive tasks. It would be surprising indeed if the workman's propensity to invent has not been diminished by the extreme specialization that characterizes the capitalist division of labor'. These considerations, we argue, can be applied when looking at the division of labour across localities. In particular we refer to the distinction between 'fragmentation' as opposed to 'specialisation' described in section 1.
19. Hamel (1991), for example, has noticed that in strategic alliances – depending on the degree of access and internalisation of new knowledge that partners can achieve by working together – there may be a relevant 'reapportionment of skills' between partners.

This uneven learning changes the relative power of actors within the alliance. Therefore, the distribution of power within economies may also be partly linked to the endogenously determined capabilities of individuals and organisations. See also Simon (1999).

20. As discussed in Wilson (Chapter 5), the corporatisation of universities is a widespread phenomenon which relates also and perhaps more prominently to so-called developed countries.

REFERENCES

Arthur, Brian W. (1990), 'Silicon Valley locational clusters: when do increasing returns imply monopoly?', *Mathematical Social Sciences* **19**(3), 235–51.

Baetjer, Howard (2000), 'Capital as embodied knowledge: some implications for the theory of economic growth', *Review of Austrian Economics* **13**(2), 147–74.

Blomström, Magnus (1986), 'Foreign investment and productive efficiency: the case of Mexico', *Journal of Industrial Economics*, **35**(1), 97–110.

Blomström, Magnus and Ari Kokko (2002), 'FDI and human capital: a research agenda', OECD technical papers no. 02/195.

Blomström, Magnus, Ari Kokko and Mario Zejan (1994), 'Host country competition and technology transfer by multinationals', *Weltwirtschaftliches Archiv, Review of World Economics*, **130**, 521–33.

Chung, W., W. Mitchell and B. Yeung (2003), 'Foreign direct investment and host country productivity: the American automotive component industry in the 1980s', *Journal of International Business Studies*, **34**(2), 199–218.

Coase, Ronald H. (1937), 'The nature of the firm', *Economica*, **4**(16), 386–405.

Cohen, Wesley M. and Daniel A. Levinthal (1989), 'Innovation and learning: the two faces of R&D', *Economic Journal*, **99**(397), 569–96.

Cowling, Keith and Roger Sugden (1994), *Beyond Capitalism. Towards a New World Economic Order*, London: Pinter Publishers.

Cowling, Keith and Roger Sugden (1998), 'The essence of the modern corporation: markets, strategic decision-making and the theory of the firm', *Manchester School*, **66**(1), 59–86.

Cusmano, Lucia (2000), 'Technology policy and co-operative R&D: the role of relational research capacity', DRUID working papers series no. 00/3, Copenhagen Business School, Department of Industrial Economics and Strategy, and Aalborg University, Department of Business Studies.

Docherty, Thomas (2008), *The English Question or Academic Freedoms*, Brighton: Sussex Academic Press.

Dosi, Giovanni (1991), 'La natura e gli effetti microeconomici del progresso tecnico', in Giovanni Zanetti (ed.), *Innovazione tecnologica e struttura produttiva*, Turin: UTET.

Dosi, Gioanni, Christopher Freeman, Richard Nelson, G. Silverberg and Luc Soete (eds) (1988), *Technical Change and Economic Theory*, London: Pinter Publishers.

Duesenberry, James S. (1949), *Income, Saving and the Theory of Consumer Behavior*, Cambridge, MA: Harvard University Press.

Dunning, John (1993), *Multinational Enterprises and the Global Economy*, Wokingham: Addison-Wesley.

Feldman, Maryann (1994), *The Geography of Innovation*, Dordrecht, Netherlands, Boston, MA, and London: Kluwer Academic.

Christopher Freeman (1991), 'The Nature of innovation and the evolution of the productive system', in OECD, *Technology and Productivity: the Challenge for Economic Policy*, Paris: OECD, pp. 303–14.

Hamel, Gary (1991), 'Competition for competence and inter-partner learning within international strategic alliances', *Strategic Management Journal*, **12**, pp. 83–103.

Haskel, Jonathan E., Sonja C. Perreira and Matthew Slaughter (2002), 'Does inward foreign direct investment boost the productivity of domestic firms?', National Bureau of Economic Research working papers series 02/8724.

Hymer, Stephen H. (1970), 'The efficiency (contradictions) of multinational corporations', *American Economic Review*, **60**(2), 441–8.

Hymer, Stephen H. (1972), 'The multinational corporation and the law of uneven development', in J.N. Bagwaty (ed.), *Economics and World Order from the 1970s to the 1990s*, London: Collier-Macmillan, pp. 113–40.

Hymer, Stephen, Burton A. Weisbrod and Harry G. Johnson (1966), 'Discussion', *American Economic Review*, **56**(1/2), 275–83.

Kern, Horst and Michael Schumann ([1984] 1991), *Das Ende der Arbeitsteilung?* Becksche, Italian translation (1991), *La fine della divisione del lavoro?*, Turin: Einaudi.

Ari Kokko (1994), 'Technology, market characteristics, and spillovers', *Journal of Development Economics*, **43**(2), 279–93.

Kondrat'ev Nicolaj D. ([1926] 1935), 'The long waves in economic life', *Review of Economic Statistics*, **17**(6), 105–15.

Lee, Jeong-Yeon and Edwin Mansfield (1996), 'Intellectual property protection and U.S. foreign direct Investment', *Review of Economics and Statistics*, **78**(2), 181–86.

Love, James (2003), 'Technology sourcing versus technology exploitation', *Applied Economics*, **35**(15), 1667–78.

Lundvall, Bengt-Åke (ed.) (1992), *National Systems of Innovation. Towards a Theory of Innovation and Interactive Learning*, London: Pinter Publishers.

Mansfield, Edwin and Anthony Romeo (1980), 'Technology transfer to overseas subsidiaries by U.S.-based firms', *Quarterly Journal of Economics*, **25**(4), 737–50.

Marglin, Stephen (1974), 'What do bosses do? The origins and functions of hierarchy in capitalist production', *Review of Radical Political Economics*, **6**(2), 60–112.

Marx, Karl ([1867] 1933), *Capital*, edited by George D.H. Cole, Bk I, Ch XIII, London: Dent.

Maskus, Keith E. and Guifang Yang (2000), 'Intellectual property rights, foreign direct investment, and competition issues in developing countries', *International Journal of Technology Management*, **19**(1–2), 22–34.

Michie, Jonathan, Christine Oughton and Pianta Mario (2002), 'Innovation and the economy', *International Review of Applied Economics* **16**(3), 253–64.

Myrdal, Gunnar (1957), *Economic Theory and Under-Developed Regions*, London: Duckworth, Italian translation: *Teoria Economica e Paesi Sottosviluppati* (1974), Milan: Feltrinelli.

Narula, Rajneesh (2003), 'Multinational firms, regional integration and globalizing markets: implications for developing countries', in Robert Devlin and Antoni Estevadeordal (eds), *Bridges for Development: Policies and Institutions for Trade and Integration*, Washington, DC: Inter-American Development Bank, Chapter 10.

Nelson, Richard R. (1998), 'The co-evolution of technology, industrial structure, and supporting institutions', in: G. Dosi, David Teece and Josef Chytry (eds), *Technology, Organization, and Competitiveness*, Oxford: Oxford University Press, pp. 319–36.

Nelson, Richard R. and Sidney G. Winter (1982), *An Evolutionary Theory of Economic Change*, Cambridge, MA: Harvard University Press.

Nolan, Peter, Dylan Sutherland and Jin Zhang (2002), 'The challenge of the global business revolution', *Contributions to Political Economy* **21**(1), 91–110.

Nurkse, Ragnar (1953), *Problems of Capital Formation in Underdeveloped Countries*, Oxford: Basil Blackwell.

Panic, M. and P.L. Joyce (1980), 'UK manufacturing industry: international integration and trade performance', *Bank of England Quarterly Bulletin*, **20**(1) (March), 42–55.

Patel, Parimal and Keith Pavitt (1998), 'Uneven (and divergent) technological accumulation among advanced countries: evidence and a framework of explanation', in Giovanni Dosi, David Teece and Josef Chytry (eds), *Technology, Organization, and Competitiveness*, Oxford: Oxford University Press, pp. 289–317.

Patel, Parimal and Keith Pavitt (2000), 'National systems of innovation under strain: the internationalisation of corporate R&D', in Ray Barrel, Geoff Meson and Mary O'Mahony (eds), *Productivity, Innovation and Economic Performance*, Cambridge: Cambridge University Press, pp. 217–35.

Patel, Pari and Modesto Vega (1999), 'Patterns of internationalisation of corporate technology: location vs. home country advantages', *Research Policy*, **28**(2–3), 145–55.

Pavitt, Keith (1984), 'Patterns of technological change: towards a taxonomy and a theory', *Research Policy*, **13**(6) 343–74.

Perez, Carlota (1983), 'Structural change and the assimilation of new technology in the economic and social system', *Futures* **15**(4), 357–75.

Pitelis, Christos N. and Roger Sugden (eds) (1991), *The Nature of the Transnational Firm*, London: Routledge.

Ricardo, David (1819), *On the Principles of Political Economy and Taxation*, London: John Murray, Ch. XXXI.

Richardson, George (1960), *Information and Investment*, Oxford: Oxford University Press.

Sacchetti, Silvia and Roger Sugden (2003), 'The governance of networks and economic power: the nature and impact of subcontracting relationships', *Journal of Economic Surveys*, **17**(5) 669–91.

Saxenian, Annalee (1994), *Regional Advantage. Culture and Competition in Silicon Valley and Route 128*, Cambridge, MA: Harvard University Press.

Schumpeter, Joseph A. ([1912] 1934), *The Theory of Economic Development*, Cambridge, MA: Harvard University Press.

Scitovsky, Tibor (1954), 'Two concepts of external economies', *Journal of Political Economy*, **62**(2), 143–51.

Silvestri, Mario (2001), 'Linee evolutive del progresso tecnico in relazione alle problematiche economiche', in: Giovanni Zanetti (ed.), *Innovazione tecnologica e struttura produttiva*, Turin: UTET, pp. 37–62.

Simon, Herbert A. (1999), 'The many shapes of knowledge', *Revue d'Economie Industrielle,* 2nd semester, (88), 23–41.

Slaughter, Mattew J. (2002), 'Skill upgrading in developing countries: has inward foreign direct investment played a role?', OECD technical papers no. 02/192.
Smith, Adam ([1776] 1976), *An Inquiry into the Nature and Causes of the Wealth of Nations*, BkI, Chicago, IL: University of Chicago Press.
Sacchetti, S. and R. Sugden (2005), 'Mental proximity: identifying networks of mutual dependence', in T. Theurl and E.C. Meyers (eds), *Strategies for Cooperation*, Aachen, Germany: Shaker Verlag, pp. 81–105.
Sugden, Roger (1997), 'Economías Multinacionales y la Ley del Desarrollo sin Equidad', *Revista de la Facultad de Ciencías Económicas y Sociales*, **3**(4), 87–109.
United Nations Conference on Trade and Development (UNCTAD) (2000), *World Investment Report 2000: Cross Border Mergers and Acquisitions and Development*, New York: United Nations.
United Nations Development Programme (UNDP) (1990), *Human Development Report 1990*, Oxford: Oxford University Press.
Veblen, Thorstein (1915), *Imperial Germany and the Industrial Revolution*, London: Macmillan.
Vernon, Raymond (1966), 'International investment and international trade in the product cycle', *Quarterly Journal of Economics*, **80**(2), 190–207.
Vernon, Raymond and W.H. Davidson (1979), 'Foreign production of technology-intensive products by U.S.-based multinational enterprises', Harvard University Graduate School of Business Administration working papers series no. 79/5.
Weber, Max ([1905] 2001), *The Protestant Ethic and the Spirit of Capitalism*, London: Routledge.
Young, Iris Marion ([1990] 1996), *Justice and the Politics of Difference*, Princeton, NJ: Princeton University Press, Italian translation (1996), *Le politiche della differenza*, Milano: Feltrinelli.
Zanetti, Giovanni (2001), 'Occupazione, progresso tecnico, globalizzazione: alcune riflessioni su dati a livello di settore e di impresa', in Giovanni Zanetti (ed.), *Produttività e occupazione: risultati di ricerche empiriche*, Milan: Franco Angeli, pp. 59–105.
Zeitlin, Maurice (1974), 'Corporate ownership and control: the large corporations and the capitalist class', *American Journal of Sociology*, **79**(5), 1073–119.

3. Harmonization, differentiation, and development: the case of intellectual property in the global trading regime

Kenneth C. Shadlen

One of the most enduring points of conflict in the global political economy is whether international economic rules and regulations should accommodate diversity or encourage harmonization of regulatory institutions. The logic of differentiation is that optimal regulatory design differs according to national needs and conditions. The logic of harmonization, in contrast, is that the transaction costs imposed by regulatory diversity impede cross-border investment and trade.

Though international economic regimes in the decades following World War II reflected both the logics of differentiation and harmonization, the former largely prevailed. Members of the General Agreement on Tariffs and Trade (GATT), for example, established basic rules and procedures for coordinating 'trade' policies, but the issue-area was narrowly defined such that countries retained virtually unlimited discretion over national regulatory institutions – even those regulatory arrangements that might affect trade flows. The logic of differentiation was even more evident with regard to developing countries: many were not members of GATT, and those that were received 'special and differential treatment' (SDT). By exempting developing countries from obligations of reciprocity, for example, poorer countries could undertake activities that were prohibited in the case of wealthier countries (Finlayson and Zacher, 1981; Tussie, 1987).

Since the early 1980s, the logic of harmonization has increasingly come to inform international economic regimes. Multilateral trade negotiations have broadened the definition of 'trade' and introduced international disciplines on a variety of regulatory practices defined as 'trade-related'. Countries are now subject to multilateral supervision not just in the form of tariffs and other non-tariff measures that have readily observable effects on trade flows (so-called 'border issues'), but also institutional arrangements that deeply affect the way national economies operate. Thus, in addition to a new set of agreements on trade in goods, the Uruguay Round, which led to the creation of the World Trade Organization (WTO), produced new agreements on subsidies, the regulation of foreign investment, trade in services (which addresses investment in financial services, telecom, and utilities infrastructure), and policies regarding intellectual property (Hoekman and Koestecki, 2001). Importantly, the dominant notion of SDT for developing countries that is incorporated into these agreements is not to set different obligations for countries at different levels of development (i.e. the logic of differentiation), but rather to establish transition periods for implementation of uniform obligations (i.e. the logic of harmonization).

The trend toward regulatory harmonization is particularly marked in intellectual property (IP). National IP laws, such as those regulating copyrights and patents, affect how private and public actors within countries absorb, adopt, and create new knowledge. Historically, differentiation has been the rule: the treatment of IP was regulated by international agreements that afforded countries a significant degree of discretion and flexibility in designing their national regimes; and national IP institutions typically corresponded to levels of economic development and innovative capacities, with wealthier countries seeking to reward those actors involved in knowledge creation and commercialisation and poorer countries seeking to promote use and dissemination of new knowledge.

Since the 1980s, international governance in IP has undergone a sea change in the direction of harmonization.[1] Reflecting a goal to universalize OECD-style IP protection, the United States and European Union worked to establish a less flexible and more enforceable set of international rules to guide national IP practices. The most important product of this campaign was the inclusion of the Agreement on Trade-Related Aspects of Intellectual Property Rights (TRIPS) as part of the new WTO. The logic of harmonization is clearly visible in TRIPS: rather than expecting variety in IP policies to correspond to levels of economic development, strong protection of IP is regarded as a driver of – and prerequisite for – economic development, and thus uniformity of treatment is deemed appropriate for countries at all income levels.

But moving toward harmonization and achieving harmonization are different matters, and it is the former that describes contemporary arrangements in IP. TRIPS constrains differentiation by establishing universal IP standards, but countries retain a wide degree of latitude with regard to how they implement the standards. Thus, while TRIPS is part of a broader phenomenon of a movement toward regulatory harmonization, a phenomenon that imposes significant constraints on areas of economic policy where countries historically had significant autonomy (UNDP, 2003; Gallagher,

2005), developing countries retain space for autonomous IP policymaking (Reichman, 1997; Correa, 2000; Watal, 2000; CIPR, 2002).

More than a decade after the introduction of TRIPS, the governance of IP, including on-going conflicts between differentiation and harmonization, remains a politically charged issue. Many developing countries, seeking to use IP policies as tools for achieving broader development goals, have sought consolidation and confirmation of the flexibilities that remain under TRIPS. The Doha Declaration on TRIPS and Public Health, for example, was the result of a coordinated campaign by developing countries to gain clear affirmation of rights set forth in TRIPS ('t Hoen, 2002; Shadlen, 2004). At the same time, many developed countries regard the standards for IP protection established by TRIPS as too weak and too easily circumvented; they are not satisfied with constraining differentiation, and instead seek to harmonize IP standards at a higher level. Thus, developed countries – especially the USA – have continued to apply direct pressure on countries to exceed their obligations under TRIPS. Most visibly, the USA secures heightened IP protection through regional and bilateral trade agreements (RBTAs), which offer market access above and beyond what is available in the WTO in exchange for IP practices that are above and beyond what is required under TRIPS (Shadlen, 2005).

In this chapter I analyse the case of IP as an on-going conflict in the global political economy over harmonization of regulatory institutions. I begin by examining the relationship between IP and development. I then show how the trend toward harmonization places new and significant restrictions on developing countries' opportunities for policy innovation in IP policy, considering the implications of harmonization for a range of issues, including late industrialization and the promotion of public health. I show that the new restrictions are most accentuated at the regional and bilateral level, where harmonization is not merely a goal sought by more developed countries but increasingly an outcome achieved. Indeed, the proliferation of RBTAs presents the most significant threat to countries' ability to use IP policies for national development purposes, as made evident by contrasting the IP provisions in the WTO with the obligations incurred by parties to RBTAs.

Before proceeding, a caveat is in order. It is obviously simplistic to discuss RBTAs as a single entity, as they exhibit considerable differences. The USA and EU, the two principal partners for such agreements, have different priorities in integrating IP into such agreements. But not only do US RBTAs differ from EU RBTAs, but all US RBTAs are not alike either.[2] Indeed, the details of the IP provisions within any given agreement are bargaining outcomes. Thus, general statements regarding IP regulations in RBTAs (US or otherwise) run the risk of distorting via oversimplification. That said, with regard to virtually any policy area, the differences between various RBTAs tend to be less than the differences between TRIPS and the RBTA closest to TRIPS, so a good deal is illuminated with simplifications. Simply stated, too much analysis of the differences *between* RBTAs without considering how the entire genre differs from TRIPS would distract our attention from the big picture.

INTELLECTUAL PROPERTY AND DEVELOPMENT: ECONOMIC TRADE-OFFS IN HISTORICAL PERSPECTIVE

Property rights are rules and regulations regarding the establishment, use, and protection of property. *Intellectual* property rights (IPRs) are a special subset on account of the distinct characteristics of the property they regulate. For the purposes of economic analysis, the most important attributes of knowledge, the underlying good that becomes 'property', are that it is imperfectly excludable and non-rivalrous. That a good is imperfectly excludable means that once an actor gains possession, the good cannot be taken away. That a good is non-rivalrous means that it can be used simultaneously by multiple people, and one person's use does not affect the amount left for anyone else.

The distinct characteristics of knowledge mean that IPRs perform different functions than property rights in other types of goods. One of the potential effects of establishing property rights over excludable and rivalrous goods is to promote optimal use. Intellectual property rights make knowledge fully excludable, at least in a sense, by providing owners with legal rights of exclusion over how actors use 'private' knowledge. The objective, however, is not to promote optimal use, but rather to stimulate supply, and the means for doing so is to *restrict* use. Indeed, a side-effect of providing rights of exclusion over non-rivalrous goods is to generate sub-optimal use. Thus, IPRs can encourage knowledge generation and commercialization by providing incentives for innovation: innovators can dedicate their time and resources toward developing new products with confidence that their ability to control distribution and use of the underlying ideas will allow them to enjoy the returns. But IPRs, by giving owners control over the distribution and marketing of the new knowledge, including the conditions under which the knowledge can be accessed and used by third parties, prevent knowledge from being disseminated and used as widely and optimally as possible. In short, IP regimes are imbued with trade-offs between stimulating knowledge *generation* and facilitating knowledge *use*.

A key point of this brief review is not simply that IP regimes perform multiple functions, to encourage both the generation and use of knowledge, but that a single set of institutions cannot maximize both objectives. That is, IP regimes have two desirable – but unavoidably conflicting – objectives, that knowledge be generated and that knowledge be used. A 'weak' IP regime that provides high incentives to use knowledge, for example by denying private rights of exclusion over some types of knowledge or simplifying third parties' access to privately owned knowledge, may not provide sufficient incentives for potential innovators. A 'strong' IP regime that gives innovators high incentives, for example by offering private rights of ownership over more types of knowledge or giving owners more rights of exclusion over knowledge, may impede use; and limited use of knowledge, in turn, can rebound negatively on future innovation, to the extent that knowledge generation is an incremental process (David, 1993; Heller and Eisenberg, 1998).

How countries prioritize the quests for generating and using knowledge has, traditionally, affected where the balance is struck in a given country at a given time. In countries with higher levels of innovative capacity, where more research and development tends to produce new knowledge, economic logic supports setting incentives to encourage and reward knowledge generation. In contrast, in countries with lower innovative capacities, where most new knowledge is that which is imported from abroad, economic logic supports setting incentives to encourage dissemination and use of new knowledge (Frischtak, 1995; Maskus and Penubarti 1995).

To gain insights on the national distribution of innovative capacities, Table 3.1 provides data on patents granted by the United States Patent and Trademark Office (USPTO), from 1997–2004. A number of points jump off the page. First, firms and organizations from the top ten developed countries account for more than 90 percent of all patents granted. Second, the US, Japan, and Germany alone account for nearly 80 percent. Third, the firms and organizations from the top ten developing and transition economies account for less than 7 percent, with more than 5 percent coming from Taiwan and South Korea. The combined total of the next eight highest ranking developing and transition countries is a mere 1.36 percent, slightly more than Italy. Whereas Table 3.1 demonstrates the concentration of knowledge-generation capacities in a handful of developed countries (with the important exceptions of Taiwan and South Korea), Table 3.2, which examines patent applications according to residency of applicant, shows the extent to which developing countries are importers and users of knowledge generated abroad. Even in the countries that the World Bank classifies as 'high income', non-resident applications overwhelm resident applications, a reflection of US dominance in this area,

Table 3.1 Patents granted by USPTO (1997–2004)

Top 10 developed countries	**Percent of total**[a]
1. USA	53.16
2. Japan	20.64
3. Germany	6.50
4. France	2.39
5. United Kingdom	2.28
6. Canada	2.09
7. Italy	1.03
8. Sweden	0.91
9. Switzerland	0.84
10. Netherlands	0.80
Sub-total	*90.65*
Top ten developing and transition countries	**Percent of total**[b]
1. Taiwan	2.87
2. South Korea	2.24
3. Israel[b]	0.57
4. Singapore[b]	0.17
5. Hong Kong	0.13
6. China	0.12
7. India	0.12
8. Russia	0.12
9. South Africa	0.07
10. Brazil	0.06
Sub-total	*6.49*

Notes:
a. Total is greater than the sum of the two sub-totals, which only include patents from the 20 countries in the table.
b. High-income country according to World Bank, but classified as 'developing country' in TRIPS.

Source: United States Patent and Trademark Office (www.uspto.gov).

but the asymmetries are even greater in the developing world. Residents account for less than 3 percent of patent applications in middle-income countries and only one-fifth of 1 percent in low-income countries.

Historically, diversity in national IP regimes – both cross-nationally and longitudinally – has corresponded to these basic national characteristics. Wealthier countries, with more innovative capacities have typically offered stronger IPRs than poorer countries. The relationship between national income and the strength of protection is illustrated by the j-curve

Table 3.2 Patent applications by residency (1997–2002)

Income levels*	(%) Non-resident applications as share of total applications
High income	82.28
Middle income	97.61
Low income	99.79

Note: * World Bank classifications

Source: World Bank, World development indicators.

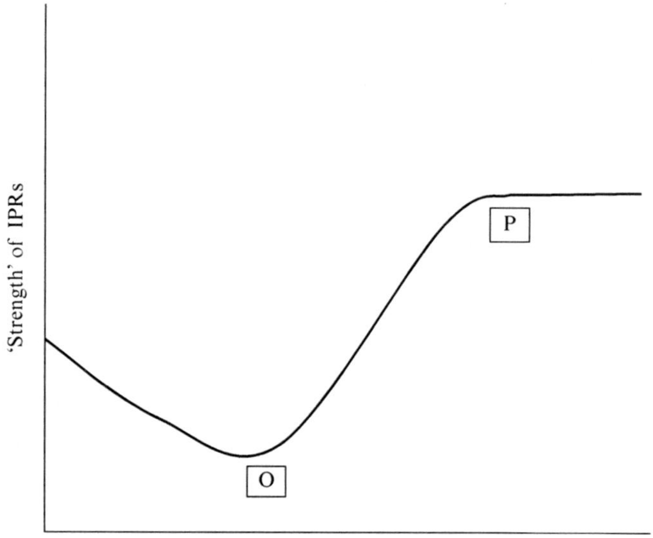

Figure 3.1 Differentiation: historical relationship between IP regime and level of development

in Figure 3.1. As countries become more industrialized and thus have greater capacity to use cutting-edge knowledge, their patent regimes tend to facilitate local firms' ability to access such knowledge (hence the dip, toward point O); later, as they develop more indigenous innovative capacities, countries' patent regimes tend to emphasize incentives for knowledge-generation (toward point P).

TRIPS makes it more difficult for developing countries to tailor their IP regimes to national conditions. Because the new standards focus primarily on establishing incentives for innovation and knowledge-generation, they limit developing countries' opportunities to design IP regimes to encourage imitation and technological learning. One concession granted to the developing countries regarded transition periods for implementation: while all countries were required to introduce national treatment and non-discrimination immediately into their existing IPR laws, developing countries had until January 2000 to bring their IPR regimes into full conformity with the WTO, the least-developed countries were given until 2006 (with the right to request extensions), and special transition periods were included for patenting pharmaceuticals and chemicals (with the least-developed countries also granted additional time). Eventually, when most transition periods are over, developed and developing countries will be subject to the same standards for IP policy, with the poorest countries still remaining exempt from some obligations – a clear reflection of the logic of harmonization. Figure 3.2 illustrates the new relationship between IP regimes and level of development, with the thick grey bars indicating the obligations established by TRIPS.

Notwithstanding the real constraints set by TRIPS, the agreement still leaves room for national variation in how countries treat IP. The borders of the upper bar in Figure 3.2 should be viewed as imprecise. The reasons for this are that the agreement is not self-executing (i.e. countries need to change their own national legislation to enter into compliance), and a number of the most important clauses are ambiguous and open-ended. When countries introduce new laws and institutions to meet their TRIPS obligations, they do so with a great deal of latitude. The result, then, is that countries may exhibit substantial variation in their patent regimes, all while being compliant with TRIPS.

In contrast to TRIPS, regional and bilateral trade agreements tend to eliminate ambiguity and establish very clearly defined obligations on how parties manage IP; and these obligations go beyond those required by even the most ambitious interpretations of TRIPS. The top line in Figure 3.2 represents the harmonization of IP institutions introduced by RBTAs.[3]

In the remainder of this chapter I contrast opportunities for IP policy innovation under TRIPS and RBTAs, with specific reference to patents. I organize the discussion around two standard limitations to the private rights conferred by patents: (1) the processes by which private rights to knowledge are obtained; (2) the extent to which private rights are subject to exceptions in the form of compulsory licenses. Within each subsection, I highlight the areas where countries retain opportunities for policy innovation despite their WTO obligations, and provide examples of how some countries have indeed introduced and retained measures that tailor IP management to local conditions and needs, all while satisfying the

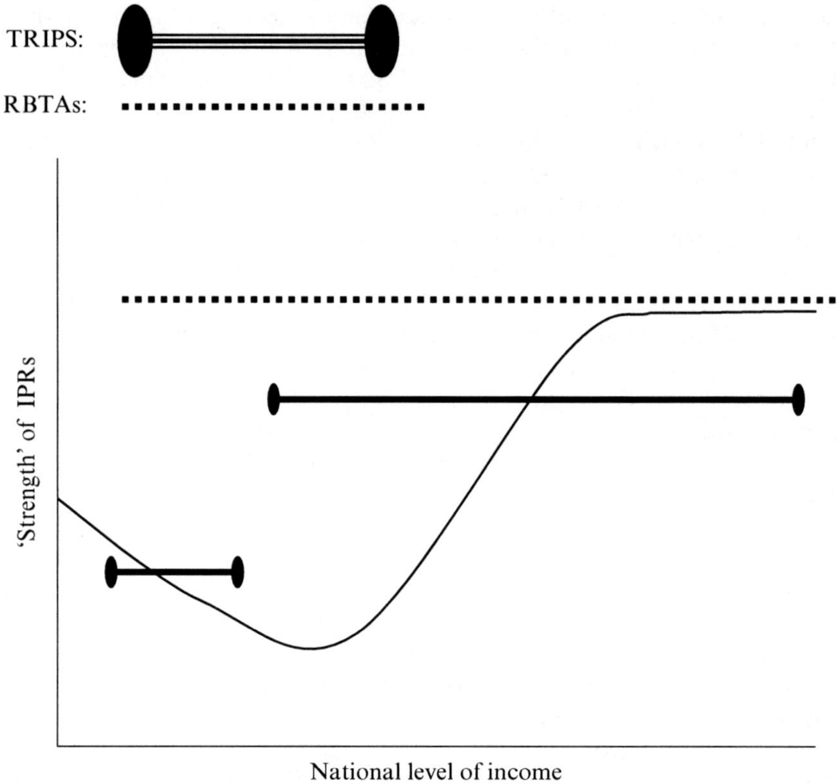

Figure 3.2 Constrained differentiation vs. harmonization: IP regimes and development levels under TRIPS and RBTAs

new TRIPS obligations.[4] Moreover, I show how these opportunities are circumscribed by RBTAs: on all dimensions of patent policy, countries that are parties to such RBTAs have significantly less autonomy in their management of IP.

WHAT KNOWLEDGE BECOMES PRIVATELY OWNED?

The first important limitation of patents is that private ownership rights are not conferred automatically upon possession of knowledge. Instead, patents are granted by the state only where applicants demonstrate that their inventions satisfy two sets of criteria: the knowledge must fall within the range of patentable subject matter; and patent examiners must determine that the knowledge is new, non-obvious, and useful. Because patent examination remains national, and with application central to the process of establishing ownership, governments delineate what knowledge can be owned privately within their territory.

Scope

The scope of a patent regime refers to the range of inventions that are eligible to be patented. This has historically been a critical feature that differentiated national approaches toward IP. Many countries refused to grant patents to certain products,[5] providing local producers' with unfettered access to foreign knowledge in key sectors has historically been a critical dimension of strategies for late industrialization, for this can facilitate local firms' abilities to adapt and build upon foreign innovations. Likewise, patents were often denied to restrain prices, encourage sharing of knowledge, and ensure that local actors (e.g. farmers) could continuously adapt to changing environmental conditions.

TRIPS reduces discretion in setting the scope of patent eligibility. Article 27 requires countries to grant patents of 20 years *in all fields of technology*.[6] Of course, any individual patent application can be denied on the standard grounds of novelty, inventiveness, and utility, and countries' national patent offices and courts retain autonomy in how they operationalize these critical concepts (as discussed below), but, in principle, countries must offer patents in all fields.

The new limitations on scope that derive from TRIPS mean that countries can no longer refuse, as a matter of policy, to issue patents to particular classes of goods, such as pharmaceutical products and chemicals. Prior to the Uruguay Round more than 40 countries did not provide any patent protection for pharmaceuticals, while many that did so issued patents only for processes and not for products (WHO, 2002, p. 15). In many developing countries, the lack of patent protection drove the growth of local pharmaceutical industries, which specialized in making generic versions of drugs – some patented in developed countries, some older drugs whose patents had expired. As of 2005, however, all but the least developed countries grant patents on pharmaceuticals and pharmo-chemical and agricultural chemicals.[7]

Examination

In addition to the question of what sort of knowledge is eligible for patenting, there is the question of how patent offices examine applications.

Under TRIPS, countries retain significant prerogatives for making private ownership of knowledge more or less simple to obtain. Most obviously, the three standard criteria for patentability – that the idea be new, non-obvious, and useful – are ambiguous terms. How these criteria are operationalized by national patent offices and legal systems affects what sorts of patents are granted. Practices established by the USPTO and EPO tend to establish some precedence in this regard, but this remains an important point of flexibility (CIPR, 2002, pp. 114–19).

Countries can set criteria for 'novelty' that makes reformulations or second uses of existing drugs ineligible for additional patents. Likewise, countries retain the freedom to determine what classifies as 'non-obvious'. India's amended Patent Act is illustrative on both accounts: the Act excludes new uses from patenting by stating that 'mere discovery' of new forms of known substances that 'do not result in the enhancement of the known efficacy' of the substances are not patentable; and the definition of 'inventive step' (used synonymously with 'non-obvious') is worded in such a way as to provide administrative and judicial officials with grounds to deny many patent applications and thus effectively narrow patent scope (Basheer, 2005).[8]

Countries also set their own definitions of an 'invention', and as such can deny patents to 'discoveries'.[9] That these are such imprecise terms certainly invites abuse, but it preserves opportunities to retain a narrow patent scope. Countries can, for example, deny patents to gene sequences, on the grounds that the technical step was a discovery of an existing entity, not an invention of something new (Demaine and Fellmeth, 2003). Restrictive definitions of invention and discovery have also been used to deny patents to computer software, under the argument that programmers are not inventing new processes but discovering or revealing underlying mathematical algorithms that are part of nature. In fact, the question of whether to grant patents to software has been extraordinarily contentious not just in the developing world but also in Europe (Haunss and Kohlmorgen, forthcoming).[10]

Countries also retain significant leeway to demand strict disclosure requirements. In exchange for exclusive rights obtained by the patent, applicants are required to make their knowledge public. The patent right sets restrictions on what can be done with the knowledge, but anyone can read and thus learn from patents. Patent applicants typically wish to reveal as little of the knowledge as possible in exchange for exclusive rights, but there is a public interest in demanding greater disclosure. The extent to which new knowledge enters the public domain and becomes available for third parties (albeit with serious restrictions on their use of the knowledge), depends on the amount and precision of disclosure that patent examiners demand.

RBTAs can erode these critical spaces for aligning policy to national conditions by exporting examination guidelines, thus removing the ambiguity that exists under TRIPS, and placing caps on the amount and type of information that patent applicants can be required to submit. DR-CAFTA, for example, defines 'novelty' in a more expansive way, exporting to all DR-CAFTA parties the more liberal meaning of 'new' that is used in the USA, where goods can pass the novelty test and be granted a patent if the knowledge has been made public within the year prior to application (Morin, 2004). RBTAs are also more likely to limit what sorts of disclosure requirements national IP offices can place on patent applicants. Again, DR-CAFTA is illustrative, for the agreement proscribes such requirements by establishing an explicit cap on the type and amount of information that countries can require from applicants (Morin, 2004). Were a Central American country to demand more information than what is necessary to repeat an invention, the country in question would most likely be in violation of its new international obligations.

It is worth underscoring that even where countries retain prerogatives on examination and disclosure requirements, using such options presents complex challenges for most developing countries. Patent applications, virtually by definition, include cutting-edge knowledge. Thus, knowing how much disclosure is 'sufficient' can be a complex task. Plus, the number of patent applications in most countries has increased astronomically since the mid-1990s, which means that national patent offices are flooded with applications on highly technical matters. For developing countries this raises an issue regarding human resource allocation. Effective patent examiners are highly-skilled and well-trained professionals with technical knowledge, normally with engineering and science backgrounds. Given that such skills are, almost by definition, in short supply in less developed countries the obvious question, then, regards the opportunity costs of deploying 'the best and brightest' as patent examiners.

Two additional areas of IP policy management must be discussed in this subsection, patent breadth and utility models. Breadth refers essentially to how many ideas (or claims) are protected by a single patent, and it, in turn, affects the terms on which follow-on innovators gain access to the patented knowledge (Merges and Nelson, 1990). Narrow patents can create opportunities for local firms and innovators to 'invent around' existing patents without being subject to litigation. Indeed, the granting of narrow patents was a key feature of Japan's postwar patent regime, one that is commonly cited as a model for late-industrializing technological followers (see discussions in Ordover, 1991; Maskus and McDaniel, 1999; Sakakibara and Bransteller, 1999; Maskus, 2000; Chang, 2002; Kumar, 2003).

Utility models (also known as 'petty patents') have shorter periods of protection than patents (seven–ten years rather then 20) for inventions that meet lower standards of inventiveness (i.e. the 'non-obvious' criterion is relaxed). They are typically made available for *incremental* inventions that build on more fundamental discoveries (Maskus, 2000: 39 and 177; CIPR, 2002: Box 1.1 and p. 175). Utility models are of particular interest in considering alternatives for IP policy in developing countries, for the degree of innovation required for protection may be more appropriate for local firms. The sorts of innovations undertaken by local firms are less likely to meet the inventiveness threshold for patentability. By granting utility models, then, developing countries can reward the smaller and incremental types of innovation that are common among local firms.

Comparative historical analysis suggests that utility models can be critically important dimensions of patent regimes. The sorts of innovations rewarded by utility models may be developmentally significant and worth encouraging, even if not strictly patentable. Analysts of the role of IPRs in East Asian development typically emphasize not just the narrow scope of patentability, for example, but also the use of utility models in Japan, Korea, and Taiwan (Kumar, 2003; Maskus and McDaniel, 1999). Indeed, a great number of developing countries narrowed the scope of patent eligibility in the postwar era, but one crucial difference that set the East Asian countries apart is that they also actively promoted utility models to encourage local firms to make adaptive and incremental innovations.[11]

Note that neither the WTO nor RBTAs address either patent breadth or utility models. Here too, however, the formal opportunities for policy innovation – though important – are difficult to exploit. Patent breadth, for example, tends not to be a function of statute, so much as of administrative and judicial practice (i.e. how patent examiners proceed, and what legal doctrines judges use in deciding infringement cases). Thus, it is 'unclear how developing countries can ensure that courts interpret claims in a narrow way, unless this is laid down in detailed guidelines, a stupendous task in itself' (Watal, 1999, p. 119, note 12). Likewise, an effective system of utility models requires significant government promotion, since many of the smaller, local firms whose innovations might qualify for this sort of protection have little familiarity with IP. Thus, despite the fact that many developing countries now offer utility models, applications tend to be low, suggesting that the systems are underutilized.

COMPULSORY LICENSES

Patent rights include exceptions to patent-holders' ability to exert control over the use of their property. One important exception is the compulsory license (CL), where the government allows a local entity (a private firm or government agency) to produce and distribute a patented good without the owner's consent. CLs have historically been part and parcel of national patent regimes, granted by countries in a wide range of situations (Reichman and Hasenzahl, 2003).

Despite efforts by the USA in the Uruguay Round to radically circumscribe their use, TRIPS continues to leave countries with a significant degree of discretion. Article 31 establishes a set of *conditions* to be met for governments to issue CLs. For example, governments must proceed on a case-by-case basis, third parties must first seek permission of the patent-holder (i.e. the CL must follow unsuccessful negotiations, though this is waived in case of national emergency), the CL must be of limited duration (and terminated when the grounds leading to the CL are no longer there), be non-exclusive, be predominantly for the domestic market, and the patent-holder should be compensated.

Operationalizing these conditions in terms of national law leaves opportunities for differentiation. For example, countries retain leeway regarding how much negotiation for a voluntary license is required before a third party can legitimately request a compulsory license from the state. Third parties must attempt to gain authorization from the patentee, and the state may only grant a compulsory license if negotiations are not successful within a 'reasonable period of time', but the determination of reasonable is left to individual countries. Likewise the requirement that 'adequate remuneration' be paid to the patentee, but countries establish their own definitions of 'adequate'. During the Uruguay Round negotiations, the USA sought to include a requirement to 'compensate the right-holder fully' (Watal, 1999, p. 114), but this language is not included in TRIPS. And in both instances, with regard to negotiations and compensation, TRIPS permits national-level interpretation and adjudication to be administrative, not necessarily judicial, which significantly increases the ease of requesting and acquiring CLs.

Beyond the issue of how countries implement these procedural obligations are the *grounds* that countries use for issuing CLs. Here it is important to emphasize that TRIPS is silent: countries can issue CLs for whatever reasons they choose. What this means is that so long as the procedural conditions – defined and operationalized locally – are met, countries establish their own grounds for issuing CLs.[12]

Developing countries' rights to issue CLs, particularly with regard to public health, were confirmed in the Doha Declaration on the TRIPS Agreement and Public Health (WTO, 2001). Paragraph 5.b., for example, affirms that 'each member has the right to grant compulsory licences and the freedom to determine the grounds upon which such licences are granted'. Thus, developing countries are only required to abide by the conditions stipulated in Article 31. Furthermore, even some of these conditions can be waived in the context of national emergencies, and paragraph 5.c. of the Doha Declaration stipulates that 'each member has the right to determine what constitutes a national emergency or other circumstances of extreme urgency. . .'.

While much of the debate over CLs has been related to issues of health and access to medicines, the relevance and importance of compulsory licensing goes beyond health. Indeed, if one contrasts contemporary debates over CLs with the previous debates over them that occurred in the 1980s (Sell, 1998, Chapter 4), it is striking how the overarching issues have changed: contemporary debates are about public health, previous debates were about the role of CLs in national strategies to promote indigenous technological advancement and industrial development. In discussing developing countries' flexibilities in this regard, then, it is worthwhile to consider the broader relevance of CLs.

By requiring patent-holding firms to manufacture their inventions locally in order to retain exclusive rights, as many countries did in the past and some (e.g. Brazil and India) continue to do under TRIPS, developing countries can encourage the transfer of non-codified, tacit knowledge that only occurs via the localization of manufacturing operations.[13] To understand the importance of the 'local working' requirement, recall two points discussed above. First, most patents are held by firms in a handful of countries, reflecting the extraordinary asymmetries in international innovative capacities (see Table 3.1). Second, patent regimes in the developing world are less about stimulating innovation than about capturing the benefits of foreign innovation through the transfer, absorbtion, and adaptation of foreign technologies. A partial condition for technological learning is that the technology is used locally. Many developing countries maintain that if patented goods are simply imported, with local use impeded by the rights of exclusion granted to foreign patent holders, technological transfer will be minimal. To the extent that governments want to improve local actors' access to and abilities to learn from foreign technologies, there may be a public interest in assuring that the technologies are used locally. One policy instrument to achieve this goal is to establish failure to produce the good locally as a ground for a CL.

Whether or not countries can establish the absence of local production as a ground for issuing a CL is an unresolved issue in TRIPS. On the one hand, the previous discussion, which indicated countries' freedoms to establish their own criteria, would suggest that such practices are permissible. On the other hand, the agreement also stipulates that patents 'shall be available and patent rights enjoyable without discrimination as to . . .

whether products are imported or locally produced' (TRIPS, Art. 27.1). The lack of certainty is illustrated by a conflict between the USA and Brazil over the latter's local working provisions. Article 68 of Brazil's 1997 industrial property law authorizes the government to issue CLs when patented goods are not produced locally after three years from the grant of the patent. The USA objected and requested a WTO hearing, accusing Brazil of being in violation of TRIPS. In June 2001, when the two countries signed a joint communiqué announcing the withdrawal of the US challenge in the WTO, they also recognized that the fundamental conflict over Article 68 remains unresolved. In the meanwhile, however, Brazil's law remains in force, and India retained a similar provision in final amendments to the Patent Act.

Before proceeding, more discussion of the US–Brazil conflict is worthwhile, for the entire affair underscores the salience of CLs as tools of industrialization and not just mechanisms to lower drug prices and promote public health. The US challenge to Brazil's IP regime has been portrayed as an attack on Brazil's globally admired public health strategy to provide universal anti-retroviral treatment for people living with HIV/AIDS and as instance where activist campaigning and negative publicity led the USA to drop its case. Both interpretations are accurate on the net, but partially misleading. With regard to the substance of the conflict, note that Brazil has two CL clauses, one for public health (Art. 71) and one for local working (Art. 68), and at precisely the same time as the conflict with the US over the latter was transpiring, Brazil was reforming the former to make CLs simpler. The USA acknowledges that Art. 71 is acceptable under TRIPS, but objects to Art. 68, which, US officials complain – explicitly – is about industrial promotion. According to the USTR's 2001 Special 301 Report on IP practices:

> [S]hould Brazil choose to compulsory license anti-retroviral AIDS drugs, it could do so under Article 71 of its patent law, which authorizes compulsory licensing to address a national health emergency, consistent with TRIPS, and which the United States is not challenging. In contrast, Article 68 – the provision under dispute – may require the compulsory licensing of any patented product, from bicycles to automobile components to golf clubs. Article 68 is unrelated to health or access to drugs, but instead is discriminating against all imported products in favor of locally produced products. In short, Article 68 is a protectionist measure intended to create jobs for Brazilian nationals. (USTR, 2001, p. 10)

Thus, at the heart of the US challenge to Brazil was a conflict over industrial strategy and developing countries' capacities to mediate their terms of integration into the international economy. In fact, as indicated in the 2000 Special 301 Report, where the USTR explained the rationale for the challenge in the first place, the US sought 'to address the concern that other countries may cite the Brazilian "local working" requirement as a justification for proposing similar legislation' (USTR, 2000, p. 7).

With regard to the 'resolution' of the case, in particular the US change of strategy, while there is no doubt that the USA received relentless criticism from all quarters for its perceived attack on Brazil's HIV/AIDS treatment program, an equally plausible explanation for why the USA dropped the case is because it feared that Brazil would win. Or more accurately, it might lose by winning. After all, the Brazilian government responded to the US challenge by pointing out various provisions in US law also discriminate require local working and thus would violate the US-favored interpretation of Art. 27.1 of TRIPS.[14] To be sure, in choosing not to press forward with its challenge the US government may have wanted to silence its critics, but it is also clear that leaving the issue unresolved in the form of a joint communiqué was preferable to losing the case and allowing a pro-industrial strategy precedent to be set.

Although developing countries that are compliant with TRIPS retain significant rights to use CLs as policy instruments, these rights are circumscribed in some RBTAs, which fuse the conditions and grounds into specific *circumstances* under which CLs can be issued. Not all RBTAs restrict the use of CLs, but the trend is to allow governments to issue CLs only as remedies for anti-competitive practices, for public, non-commercial use, and in times of national emergency or 'other circumstances of extreme urgency'.[15] And even then, patent-holders are due 'reasonable and entire compensation' (much tighter and stronger language than in TRIPS). The precise language varies across RBTAs, with the strongest restrictions in the US–Singapore agreement. The language on CLs in the US–Morocco RBTA, in contrast, is much weaker, a fact that drew the wrath of the US industry group that advises USTR on the IP aspects of trade policy, the Industry Functional Advisory Committee on Intellectual Property Rights for Trade Policy (IFAC-3).[16]

With regard to local working requirements, the USA gets around the problem revealed by the conflict with Brazil by putting more explicit restrictions on CLs directly into the relevant section of the RBTA. To explain, Brazil's local working requirement is clearly within its CL rights under TRIPS Art. 31, but allegedly violates its non-discrimination obligations under TRIPS Art 27. As indicated, one of the problems for the USA was that it too is potentially in violation of TRIPS Art 27. RBTAs eliminate this problem by proscribing such practices directly. By explicitly listing the limited and exclusive conditions under which CLs can be granted, local working requirements of the sorts found in Brazil's and India's patent regimes are prohibited.

In sum, RBTAs tend to pick up where the WTO leaves off in terms of limiting developing countries' abilities to deploy what historically have been standard tools to regulate patent holders. Neither sort of agreement prohibits CLs, but some RBTAs establish clear and unequivocal biases against their use – biases that are significantly stronger than in TRIPS. Whereas TRIPS allows governments to issue CLs on any grounds provided they take certain measures, some RBTAs prohibit governments from issuing CLs except in very strictly and tightly defined circumstances.

CONCLUSION

That opportunities for national differentiation in IP policy are reduced under TRIPS and RBTAs is illustrated with reference to the axes of variation utilized in the previous sections: countries must implement measures to make establishment of private rights over knowledge easier to obtain, and the subsequent private rights must be more absolute. Whereas countries could previously deny patents to certain types of inventions in order to encourage reverse-engineering and lower the barriers to entry in technologically-intensive sectors, now countries must offer patents in virtually all fields. And whereas countries could make enjoyment of the monopoly rights conferred by patents conditional upon local production or licensing and transferring technology to local users, governments are now limited in how they regulate patent-holders.

Yet a number of key points should stand out from the preceding contrast between space for IP policy available under the WTO and under RBTAs. First, one need not be an enthusiast of TRIPS to acknowledge that developing countries retain opportunities for policy innovation in the field of IP. To point out that TRIPS 'merely' introduces 'constrained differentiation' rather than harmonization is not to defend TRIPS. The constraints imposed by TRIPS are serious, and many important policy instruments used in the past are now prohibited. TRIPS does usher in a new environment for IP management that has rightfully caused a great deal of consternation among analysts of IP and development (May, 2000; Drahos and Braithwaite, 2002; UNDP, 2003). But governments can create TRIPS-compatible patent regimes that, by facilitating use and being geared toward adaptation and learning, may be appropriate for late development. The second key point is that these opportunities are radically reduced – if not eliminated – by RBTAs, which achieve outcomes closer to harmonization. On both dimensions used to assess IP policy – governments' abilities to determine which knowledge becomes private property and to provide for exceptions to patent-holders' exclusive rights – RBTAs place more burdensome obligations on developing countries.

The implications for late development of harmonization – in contrast to constrained differentiation – are profound. Whereas TRIPS represents a worrying step toward the danger zone, RBTAs cross over the line.[17] To understand why, it is worth revisiting the foundations of IP and development discussed earlier in this chapter. IP regimes constitute trade-offs between incentives for knowledge generation and incentives for knowledge use; and because the same set of institutions are not likely to maximize both functions, differentiation allows countries to tailor their IP regimes to national conditions and development objectives. Harmonization, in contrast, creates a world in which all countries adopt IP regimes that are designed to encourage knowledge *generation*. This change may, perhaps, lead to more technological generation at a aggregate global level – and by lowering transaction costs may make the global economy more efficient and produce aggregate welfare gains.

Yet harmonization has serious implications for development – for late development is and always has been based on borrowing more advanced knowledge and technology. By definition, late-developing countries are not at the knowledge frontier; they need to adapt foreign innovations for local use. Late development is about learning and adaptation: catching-up in the global economy does not occur via technological breakthrough so much as technological imitation and adaptation. Indeed, one of the defining principles of 'lateness' is the imperative to industrialize via learning, by borrowing and improving on technologies already developed by experienced firms in more advanced economies (Amsden, 2001). Harmonization makes this more difficult. The threat to developing countries, then, is not only in terms of resource transfers, in that net users of knowledge will have to pay more to net producers of knowledge for access to technological innovations, but that regulatory harmonization in IP may fundamentally block technological progress – and thus development.

ACKNOWLEDGEMENT

This chapter was made possible by the Globalization and Sustainable Development Program at the Global Development and Environment (GDAE) Institute at Tufts University.

NOTES

1. See, among others, Drahos (1995); Ryan (1998); Sell (1998, 2003); May (2000); Matthews (2002); May and Sell (2005); Shadlen et al. (2005).

2. In the Americas, the USA has agreements (in force or awaiting ratification) with Chile, Mexico and Canada, five countries of Central America plus the Dominican Republic, Colombia, Panama, and Peru. Plus, of course, there is the hemispheric Free Trade Agreement of the Americas, which would include 34 countries (all the sovereign states of the Americas with the exception of Cuba). Outside the Americas, the list of RBTAs that are either completed or in the process of negotiation includes (by region), the Southern African Customs Union (negotiations suspended); Bahrain, Jordan, Morocco, and Oman (also Israel; the US–Israeli agreement does not include IP provisions); Australia, Malaysia, Singapore, South Korea, and Thailand. See www.ustr.gov/Trade_Agreements/Section_Index.html.
3. The line in Figure 3.2 is placed above the earlier curve, for RBTAs do not just harmonize at the level of the wealthier country (e.g. the USA), but rather introduce new obligations on both parties.
4. Of course, many countries do not take advantage of these remaining opportunities. Explaining this underutilization is the subject of other research.
5. In the 1800s and early 1900s many countries did not grant patents at all, and many did so only to nationals. See Machlup and Penrose (1950), Schiff (1971), Chang (2002).
6. This article introduces a new definition of the term 'non-discrimination', no longer referring to countries' practices vis-à-vis other countries but rather toward economic sectors.
7. Countries that did not previously grant patents on pharmaceuticals and agricultural chemicals were given until 2005 to begin doing so.
8. After the Swiss pharmaceutical firm Novartis's application for a patent on 'Glivec' (the brand name for its anti-leukemia drug based on the molecule imatinib mesylate) was rejected by the Indian Patent Office in Chennai, on grounds on non-efficacy over existing and known substances, Novartis (unsuccessfully) challenged the TRIPS compatibility of Section 3.d in the Indian courts.
9. The word 'invention', one of the cornerstones of IP, is not defined in TRIPS.
10. Note, however, that not patenting software does not exempt countries from their obligations to provide copyright protection to software as a form of artistic expression. This is a firm and immutable obligation – albeit a new one – and an area where the USA exerts considerable pressure (Shadlen et al., 2005).
11. Contrasting India with the rapidly developing countries of East Asia, Kumar (2003) attributes the development of India's large and robust domestic pharmaceutical industry to the decision to make drugs ineligible for product patents (an uncontroversial point made by many), but also attributes the comparatively poor performance of the domestic mechanical engineering industries to the absence of utility models.
12. One place grounds are mentioned explicitly is in Art 31.k, which addresses CLs to remedy anti-competitive practices – and this clause *suspends* some of the aforementioned conditions (e.g. prior negotiations are not necessary and the CL does not need to be 'predominantly' for domestic use).
13. Another policy example is establishing restrictive licensing arrangements as a ground for a CL (as Taiwan does), which can help local firms gain access to patented knowledge on better terms.
14. This is a reference to the Bayh–Dole Act, which facilitates patenting of research generated through public funding.
15. Of course, this allows the USA (and other parties) to challenge whether or not countries are experiencing emergencies. Recall that the language of the Doha Declaration, in which countries make their own determinations regarding national emergencies, is not relevant in RBTAs.
16. 'IFAC-3 notes that the [US–Morocco RBTA] fails to include explicit restrictions on a country's authority to grant compulsory licenses to situations that are needed to remedy anti-trust violations; national emergencies or other circumstances of extreme urgency; and to govern situations of public non-commercial use. IFAC-3 believes that it is critical that future FTAs include these compulsory licensing restrictions, which were found in the Singapore FTA' (IFAC-3, 2004, p. 14).

17. In Shadlen (2005) I explain more generally the developmental implications of the trade-offs involved in RBTAs.

REFERENCES

Amsden, Alice (2001), *The Rise of 'The Rest': Challenges to the West from Late-Industrializing Economies*, New York: Oxford University Press.

Basheer, Shamnad (2005), 'India's tryst with TRIPS: the patents (amendment) Act 2005', *Indian Journal of Law and Technology*, **1**, 15–46.

Basheer, Shammad and Prashant Reddy (2008), '"Ducking" TRIPS in India: a saga involving Novartis and the legality of Section 3(d)', *National Law School of India Review*, **20**(2), 131–55.

Chang, Ha Joon (2002), *Kicking Away the Ladder: Development Strategy in Historical Perspective*, London: Anthem Press.

Commission on Intellectual Property Rights (CIPR) (2002), *Integrating Intellectual Property Rights and Development Policy*, London: CIPR.

Correa, Carlos (2000), *Intellectual Property Rights, the WTO and Developing Countries: The TRIPS Agreement and Policy Options*, London and New York: Zed Books.

David, Paul A. (1993), 'Intellectual property institutions and the Panda's thumb: patents, copyrights, and trade secrets in economic theory and history', in Mitchel B. Wallerstein, Mary Ellen Mogee and Roberta Schoen (eds), *Global Dimensions of Intellectual Property Rights in Science and Technology*, Washington, DC: National Academy Press, pp. 19–61.

Demaine, Linda J. and Aaron X. Fellmeth (2003), 'Patent law: natural substances and patentable inventions', *Science*, **300**, 1375–6.

Drahos, Peter (1995), 'Global property rights in information: the story of TRIPS at the GATT', *Prometheus*, **13**, 6–19.

Drahos, Peter, and John Braithwaite (2002), *Information Feudalism: Who Owns The Knowledge Economy?*, London: Earthscan.

Finlayson, Jock A. and Mark W. Zacher (1981), 'The GATT and the regulation of trade barriers: regime dynamics and functions', *International Organization*, **35**(4) (Autumn), 561–602.

Frischtak, Claudio (1995), 'Harmonization versus differentiation in intellectual property right regimes', *International Journal of Technology Management, Special Issue on the Management of International Intellectual Property*, **10**, 200–13.

Gallagher, Kevin P. (ed.) (2005), *Putting Development First: The Importance of Policy Space in the WTO and International Financial Institutions*, London: Zed Books.

Haunss, Sebastian and Lars Kohlmorgen (forthcoming), 'Lobbying or politics? Political claims making in IP conflicts', in Sebastian Haunss and Kenneth C. Shalden (eds), *The Politics of Intellectual Property: Contestation Over the Ownership, Use and Control of Knowledge and Information*, Cheltenham, UK and Northampton, MA, USA: Edward Elgar.

Heller, Michael A. and Rebecca S. Eisenberg (1998), 'Can patents deter innovation? The anticommons in biomedical research', *Science*, **280**, 698–701.

Hoekman, Bernard M. and Michel M. Kostecki (2001), *The Political Economy of*

the World Trading System: The WTO and Beyond, 2nd edn, New York: Oxford University Press.
Industry Functional Advisory Committee on Intellectual Property Rights for Trade Policy Matters (IFAC-3) (2004), 'The U.S.–Morocco Free Trade Agreement (FTA): the intellectual property provisions', report of the Industry Functional Advisory Committee on Intellectual Property Rights for Trade Policy Matters, 6 April, accessed at http://permanent.access.gpo.gov/websites/www.ustr.gov/assets/Trade_Agreements/Bilateral/Morocco_FTA/reports/asset_upload_file164_3139.pdf.
Kumar, Nagesh (2003), 'Intellectual property rights, technology and economic development: experiences of Asian countries', *Economic and Political Weekly*, 18 January, 209–25.
Machlup, Fritz and Edith Penrose (1950), 'The patent controversy in the nineteenth century', *Journal of Economic History*, **10**, 1–29.
Maskus, Keith (1997), 'Implications of regional and multilateral agreements for intellectual property rights', *World Economy*, **20**, 681–94.
Maskus, Keith (2000), *Intellectual Property Rights in the Global Economy*, Washington, DC: Institute for International Economics.
Maskus, Keith E. and Christine McDaniel (1999), 'Impacts of the Japanese patent system on productivity growth', *Japan and the World Economy*, **11**, 557–74.
Maskus, Keith E. and Mohan Penubarti (1995), 'How trade related are intellectual property rights?', *Journal of International Economics*, **39**, 227–48.
Matthews, Duncan (2002), *Globalising Intellectual Property Rights: The TRIPs Agreement*, London: Routledge.
May, Christopher (2000), *A Global Political Economy of Intellectual Property Rights: The New Enclosures?*, London and New York: Routledge.
May, Christopher and Susan K. Sell (2005), *Intellectual Property Rights: A Critical History*, Boulder, CO: Lynne Rienner Publishers.
Merges, Robert P. and Richard R. Nelson (1990), 'On the complex economics of patent scope', *Columbia Law Review*, **90**, 839–916.
Morin, Jean-Frédéric (2004), 'The future of patentability in international law according to DR-CAFTA', *Bridges Monthly*, March 2004, pp. 14–16.
Ordover, Janusz A. (1991), 'A patent system for both diffusion and exclusion', *Journal of Economic Perspectives*, **5**, 43–60.
Reichman, J.H. (1997), 'From free riders to fair followers: global competition under the TRIPs Agreement', *New York University Journal of International Law and Politics*, **29**, 11–93.
Reichman, Jerome H. and Catherine Hasenzahl (2003), 'Non-voluntary licensing of patented inventions: historical perspective, legal framework under TRIPS, and an overview of the practice in Canada and the USA', UNCTAD-ICTSD project on IPRs and Sustainable Development issue paper no. 5, accessed at www.ictsd.org/pubs/ictsd_series/iprs/CS_reichman_hasenzahl.pdf.
Ryan, Michael (1998), *Knowledge Diplomacy: Global Competition and the Politics of Intellectual Property*, Washington, DC: Brookings Institution Press.
Sakakibara, Mariko and Lee Branstetter (1999), 'Do stronger patents induce more innovation? Evidence from the 1988 Japanese patent law reforms', NBER working paper no. 7066.
Schiff, Eric (1971), *Industrialization without National Patents*, Princeton, NJ: Princeton University Press.

Sell, Susan K. (1998), *Power and Ideas: North-South Politics of Intellectual Property and Antitrust*, Albany, NY: State University of New York Press.
Sell, Susan K. (2003), *Private Power, Public Law: The Globalization of Intellectual Property Rights*, London: Cambridge University Press.
Shadlen, Kenneth C. (2004), 'Patents and pills, power and procedure: the North–South politics of public health in the WTO', *Studies in Comparative International Development*, **39**, 76–108.
Shadlen, Kenneth C. (2005), 'Exchanging development for market access? Deep integration and industrial policy under multilateral and regional-bilateral trade agreements', *Review of International Political Economy*, **12**, 751–75.
Shadlen, Kenneth C. (2008), 'Globalisation, power and integration: the political economy of regional and bilateral trade agreements in the Americas', *Journal of Development Studies*, **44**(1), 1–20.
Shadlen, Kenneth, C., Andrew Schrank and Marcus Kurtz (2005), 'The political economy of intellectual property protection: the case of software', *International Studies Quarterly*, **49**, 45–71.
't Hoen, Ellen (2002), 'TRIPS, pharmaceutical patents, and access to essential medicines: a long way from Seattle to Doha', *Chicago Journal of International Law*, **3** (Spring), 27–46.
Tussie, Diana (1987), *The Less Developed Countries and the World Trading System: A Challenge to the GATT*, New York: St. Martin's Press.
United Nations Development Programme (UNDP) (2003), *Making Global Trade Work for People*, London: Earthscan.
United States Trade Representative (USTR) (2000), '2000 special 301 report', Washington, DC: USTR.
USTR (2001), '2001 Special 301 Report', Washington, DC: USTR.
USTR (2003), 'Free trade with Central America: summary of the U.S.–Central America Free Trade Agreement', Washington, DC: USTR.
USTR (2004), 'U.S.–Morocco Free Trade Agreement: access to medicines', Washington, DC: United States Trade Representative.
van Bael, Ivo and Jean François Bellis (1990), *Anti-dumping and other Trade Protection Laws of the EEC*, 2nd edn, Bicester: CCH Editions.
Vivas-Eugui, David (2003), 'Regional and bilateral agreements and a TRIPS-plus world: the Free Trade Area of the Americas (FTAA)', TRIPS issue papers 1, Geneva: Quaker United Nations Office (QUNO), accessed at http://geneva.quno.info/pdf/FTAA%20(A4).pdf.
Watal, Jayashree (1999), 'Implementing the TRIPS Agreement on patents: optimal legislative strategies for developing countries', in Owen Lippert (ed.), *Competitive Strategies for the Protection of Intellectual Property*, Vancouver, BC: Fraser Institute, pp. 105–24.
Watal, Jayashree (2000), *Intellectual Property Rights in the World Trade Organization: The Way Forward for Developing Countries*, New Dehli: Oxford University Press, India.
World Health Organization (WHO) (2002), 'Network for monitoring the impact of globalization and TRIPS on access to medicines', EDM health economics and drugs series no. 11, WHO/EDM/PAR/2002.1.
World Trade Organization (WTO) (2001), 'Doha declaration on the TRIPS Agreement and public health', WT/MIN(01)/DEC/W/2 14 November, accessed at www.wto.org/english/thewto_e/minist_e/min01_e/mindecl_trips_e.htm.

4. Knowledgeable regions, Jacobian clusters and green innovation

Philip Cooke

INTRODUCTION

As is well-known, for those interested in evolutionary economic geography, Schumpeter left almost no regional or spatial analysis of economic phenomena. From the evolutionary economic geography and policy viewpoints, this is clearly disappointing. His two brief allusions are highly time–space specific. The first concerns Schumpeter's fifth form of innovation, which he designated 'railroadization' – the phenomenon by which US agricultural lands were opened up to markets by infrastructural investments, not only in railroads but farms, grain silos and even agricultural manuals that the railroad companies of the western USA had printed so that pioneers accessing cheap land on the plains would know how to farm that land. This 'regional evolution' of land and markets was, rightly, considered an *externalized* organizational innovation as compared with the *internalized* organizational innovation of a corporation adopting new management methods that gave it an, albeit temporary, competitive edge (Schumpeter, 1975). The second allusion is even briefer where Schumpeter mentions innovation such as the department store only being feasible in the large city due to the level of demand required to sustain such an innovation. Hence the city is seen as having some economic specificity from its scale attributes, but Schumpeter says nothing about the dynamics of the entailed processes (Andersen, 1994, 2007).

However, this chapter suggests that bemoaning Schumpeterian neglect of the spatial dimension may be misplaced. His category of innovation by railroadization helps understanding of regional innovation in which clusters 'mutate' through a Jacobian (after Jane Jacobs, 1969) related variety operating at regional level in places like California, North Jutland (Denmark) and Wales (UK). Regional innovation through cluster mutation is illuminated by the interest of this chapter in 'green' innovation, which is pronounced in those regions. These are the only regions to have been examined from a 'green innovation' perspective thus far to this author's knowledge. A possible reason for the illuminative aspect of taking a green perspective is that green innovation (such as the burgeoning *Cleantech* industry) displays a high degree of innovation convergence across fields like information and communication technology (ICT), nanotechnology, biotechnology, agro-food, health, environment, energy, production and materials management and waste treatment. Thus innovation occurs laterally among distinctive parts of what may be described as an innovation *platform*. Other regions for which Jacobian clustering is probably true are those of the Third Italy, which has been studied from this perspective by Boschma (2005) also from an evolutionary economic geography viewpoint. He found that apparently different industrial districts displayed 'related variety' in their *engineering* competences and associated high lateral absorptive capacity towards innovations emanating from neighbouring industries and clusters. No claim is made here for the ubiquity of this process, on the contrary, Jacobian cluster regions are probably not in the majority. But where they exist they can be propulsive in relation to national economies or aspects of them. To that extent they make a contribution to understanding of regional and national unequal development between wealth and poverty, the issue that has animated economics since Adam Smith.

Having held out the promise of a Neo-Schumpeterian theory of regional evolution, that aspiration has to be severely qualified. For a more truly evolutionary theory of spatial dynamics we have to turn to the mid-twentieth century inheritors of Veblen's concept of 'cumulative causation'. A variant of the biblical 'Matthew principle' of 'to those that have, more shall be given' this profoundly disequilibrium perspective contains the missing dynamic element by virtue of Myrdal's (1957) elaboration upon the various 'backwash' and 'spread' effects associated with regional evolution. Spread effects, on occasions, caused the dynamic element to seek to accommodate growth beyond its original boundaries. Backwash effects sucked back temporary gains made by competing locations to the larger, predominating accumulating entity, such as a strong city or regional economy. Observations of static relationships in the spatial evolution of the 'knowledge economy' have led to the preliminary postulation of a knowledge capabilities theory of regional evolution based on the distinctive distribution of two key components of the knowledge economy labour market (Cooke, 2007). Foremost here are, first, the knowledge intensive business services (KIBS) such as finance, research, media, software and so on. While second, high technology manufacturing is a mainstay of the knowledge economy in computer and communications hardware, aerospace and biotechnology inter alia. Empirical observation of the static picture for the EU strongly suggested an urban–regional split between the locations of these. The former predominate particularly in primate cities (such as the major financial centre, sometimes, but not always, combining capital city administrative functions); the latter predominate in specialized satellite towns, often with appropriate knowledge centres like national research institutes or universities centred in them. This theory, in brief, is consistent with Myrdal–Hirschman[1] theses about 'cumulative causation' and metropolitan regional concentration of knowledge economy activities (Cooke, 2002). But as noted, the static picture merely hinted at the dynamism explicit in the idea of cumulative causation, which remained to be tested. The first such contemporary test was accomplished by access to and analysis of special runs of Israeli data in a dynamic perspective (Cooke and Schwartz, 2008). This short chapter relies on those findings and then explores innovation outside primate cities where knowledge-intensive business services (KIBS) thrive to explore innovation at some distance from big cities altogether. Nevertheless Myrdal-Hirschman models postulate innovation as capable of occurring there because such cities tend not to localize high technology manufacturing (HTM). To that we would add that they do not tend to have functional regional innovation systems and they are often the home of specialized rather than related variety clusters. The next sections take these insights and exemplifies this by reference to some 'cumulative causation' peculiarities of innovative industry regions. Those selected display innovative convergence among high technology sectors to contribute to cleaner manufacturing, food and energy production – the so-called *Cleantech* sectors, respectively (Cooke, 2008b).

THE KNOWLEDGE ECONOMY: WHAT IS IT?

It is important to say straightforwardly that the deployment of *knowledge* in economic affairs is not a new thing. Making a fire is clearly a knowledgeable and, in the deep past, powerful, knowledge-based skill, as the Prometheus myth testifies. Hunting, farming, smelting copper, bronze and iron and later, steel, are knowledge-based activities. In turn these knowledges became the basis for science and its application in early industrial technology. From coal mining grew coal tar production, the origin of the German dyestuffs industry whose aniline products led to branching into pharmacology, the (re-) discovery by the Bayer Corporation of aspirin and the birth of modern pharmaceuticals. This industry is now shifting from its synthetic chemistry origins into post-genomics and other variants of molecular biology and the science-based biotechnologies of the future.

Thus the underlying idea of a knowledge economy refers to specific assets that consist in knowledge 'how to', 'who to' and 'what to' deploy to create value. It is an active economic practice rather than a passive information space, upon which it nevertheless depends, but in ways that express value through the scarcity of 'knowledgeable' expertise. Manuel Castells (1996) speaks of the knowledge economy being one in which productivity derives from the interaction of knowledge upon knowledge rather than upon raw materials. Nonetheless, it is wrong to dismiss traditional or 'old economy' economic activity as not belonging to the knowledge economy, as for example the OECD does. It places the food industry in the low technology category, although Smith (2000) shows it to be heavily *scientific* knowledge using more than producing. Nevertheless, while 'functional foods' occupy probably a smaller segment of total food sales than the competing organic food they are both intensive utilizers of biotechnology. Surprisingly perhaps, non-genetically modified knowledge is used in organic breeding via DNA or molecular 'markers' that speed-up the breeding process for plants and animals. Thus we may also usefully speak of 'pure' and 'applied' knowledge economy activity; the first captured in genomics, software and, for example, 'futures' or derivatives trading in financial services, or conceptual art. The second is in many other sectors that conduct or use R&D even though it is applied to, for example, food production, fashion design, or fire insurance.

This allows the following policy inference to be drawn from the application of Myrdal–Hirschman models: it is unlikely that regions aspiring to evolve upwardly in economic accomplishment will be able to do so if they concentrate on developing a full portfolio of KIBS since these are more attracted to large, even primate, cities. That is not to say that some KIBS are not needed, perhaps consultancy, management accounting, venture capital (attracted by university spinouts, incubators, etc.), software and, above all, research both private and public. These latter may occasionally involve fairly mundane economic activity support for sectors such as agro-food, nowadays innovating in such areas as 'functional food' (food biotechnology), organic food (utilizing biotechnology for enhanced crop and animal breeding utilizing 'molecular markers') and of course renewable energy from biofuels. In other words KIBS can be important specialized supports in knowledge economy regions with high-tech manufacturing (HTM) and medium-tech manufacturing (MTM) requiring also high-quality precision production that is difficult to outsource globally. Jena, for example, with its universities, polytechnic, research institutes and specialist opto-electronic corporations and spinouts has the current appearance of a relatively small-scale innovation system at sub-regional level. It has some similarities with some of the induced local innovation systems promoted in Sweden by the state innovation systems development agency VINNOVA through its Vinnväxt programme. This is an exemplification of regional innovation systems thinking translated from the academic arena into the policy field. The idea of regional innovation systems came from integrating a growing literature on networks of innovation at regional level with another on innovation policies at regional level, as follows.

REGIONAL INNOVATION SYSTEMS: INTEGRATING REGIONAL NETWORKS AND REGIONAL INNOVATION POLICY

The enthusiasm for studying networks arose in a context of manifest decline in the co-ordinating capabilities of states and markets regarding leading edge research and innovation, which subsequent data (e.g. Chesbrough, 2003) show set in from approximately 1991. But if the central state had become as debilitated as many large private corporations were to become regarding the lack of productivity from their large budgetary allocations to research and development (R&D), the 'regional state' seemed from empirical reportage of the kind discussed above to be on the rise.

A parallel strand of research had evolved, which focused on regional innovation policy (e.g. Antonelli and Momigliano, 1981; Cooke, 1985). Thus the connecting concept of 'regional innovation systems' evolved from this even earlier thinking about 'regional innovation policy', in relation to 'regional innovation networks' (the 'systems view of planning' intruding again). This happened in two publications, the more widely-cited one being less theoretically and empirically rich than the almost totally uncited one. The difference between Cooke (1992) and (1993) lies in the absence of any bibliographical influence from the 'innovation systems' literature in the 1992 paper, which thus has purer lineaments to economic geography. Contrariwise, the 1993 paper which shows the author had by then read Lundvall's (1988) contribution on 'innovation as an interactive process' to Dosi et al. (1988) and was also influenced by Johansson (1991) and Grabher (1991) in probably the first proper book on regional development from a 'network regions' perspective (Bergman et al., 1991).

It seemed necessary to place these distinctive 'network and policy' concepts in relation to each other in a layered model. So, the innovation policy dimension evolved conceptually into the idea of a subsystem supporting with knowledge and resources the innovative firms in their networks. These formed a 'superstructural' subsystem dealing with actual innovation 'near market'. As we have seen, they had been spoken of as carrying out 'networking' with each other, not only laterally in alliances or partnerships and vertically in sometimes partly localised supply chains but also with the innovation policy and knowledge generation subsystem (Meyer-Krahmer, 1990; Cooke et al., 1991; Malecki, 1991; Rothwell and Dodgson 1991). So these also had subsystem characteristics related to the governance of innovation support. Each subsystem was also seen to interact with global, national and other regional innovation actors, and even through technological or sectoral systems of innovation. Open systems ruled.

Over the years the RIS framework has been analysed in terms of many different 'varieties of innovation' relating to localized, networked and hierarchical innovation 'governance' systems. Third Italy, Baden-Württemberg and French innovative regions exemplified each, respectively. Correspondingly, the 'exploitation' subsystem of firms, in the main, could be dominated by large firms or oligopolies – even foreign ones as with the Asian transplants to Wales in the 1980s and 1990s. Other regions, like Catalonia had a mix of large (SEAT automobiles) and small and medium-sized enterprise (SME) 'district' type innovation relations, while other places might have innovation regimes in which only small, entrepreneurial firms predominated, as in places with observable 'industrial districts', not only Third Italy but also some newer technology 'clusters'. Later still, these, more entrepreneurial SME systems, living by venture capital and exploitation of public research from universities, could be differentiated further as 'entrepreneurial' (ERIS), market-led systems, compared with those, especially in Europe, where they were more 'institutional' (IRIS) where state support was pronounced and 'entrepreneurship' was less advanced (Cooke, 2004).

RECENT ADVANCES IN RIS RESEARCH

One of the most interesting research areas opened up in RIS research in the recent past concerns, once again, the insights of Jane Jacobs (1969) and can be referred to as addressing the challenging issue of 'cluster emergence'. In particular by examining the emergence of a number of 'green clusters' on a regional canvas, we see emphasis in 'green innovation' upon technological convergence among diverse industries. These include biotechnology, information technology and nanotechnology (but not limited to these high-tech activities) and among them we also see a process of cluster 'species mutation'. Of particular fascination here is that some regions have the capability to mutate relatively rapidly many 'Jacobian' clusters – so-called because although different they display evolutionary characteristics of 'related variety' (Boschma, 2005). A clear definition is called for here to denote this new concept. The key is the evolutionary concept of variety, whereby some new combinations of entrepreneurial and innovative opportunity might present themselves in geographically proximate space. This would arise from the mixture of knowledge spillovers and rather high absorptive capacity among neighbouring economic activities.

Hence (Jacobian) variety is both a context and an 'evolutionary fuel' for cluster emergence as long as there is not too much cognitive dissonance or distance between neighbouring economic activities. Hence Jacobian clusters emerge from new combinations of knowledge cross-fertilizing among, for example, high technology activities like biotechnology and information technology that may be foundations for a new clean technology cluster that adopts and adapts elements from both. But, for example, new combinations from agro-food and automotive industries that are historically not that close in technical terms may also arise if the new combination being sought is biodiesel or bioethanol. This is because adjustments in breeeding of plants may have to be made if negative effects on engine performance cannot be made by the automotive side of the equation. Hence related variety is not fixed to sectoral relatedness but also embodies particular and contextuated technological convergences. In that respect it is much harder to predict cross-fertilization in the latter than the former case. But in any case, Jacobian variety rests not within but among the clusters according to this line of reasoning. Moreover it is likely to occur in the relative geographical proximity of regions. In what follows, empirical evidence is provided of regional evolution through innovation of differing intensities 'mutating' through processes of knowledge search and selection that happen to give rise to successive clustering phenomena in regional 'platforms' of related economic variety.

Jacobian Clusters

One such region is Northern California whose ICT, biotechnology and clean technology clusters overlap in proximity to San Francisco but also near various agro-food clusters like wine in the Napa, Sonoma and Russian River valleys and varieties of horticulture in the San Joaquin and Sacramento river valleys (see Figure 4.1). But notice, Figure 4.1 also shows Southern California having prominent Jacobian clusters in Los Angeles and San Diego (Cooke, 2008a). The content of Figure 4.1 is drawn entirely from secondary evidence supplied by the numerous studies of clusters in California as published in Saxenian (1994); Porter (1998); Simard and West (2003); Guthman (2005); Scott (2006); Cooke (2007). North Jutland in Denmark is another such region, as apparently is Wales in the UK, as we shall see. North Jutland's economy is the global centre of the wind turbine production industry whose profile and evolutionary trajectory was a key beneficiary from the outset of varieties of innovation. As will be shown, this

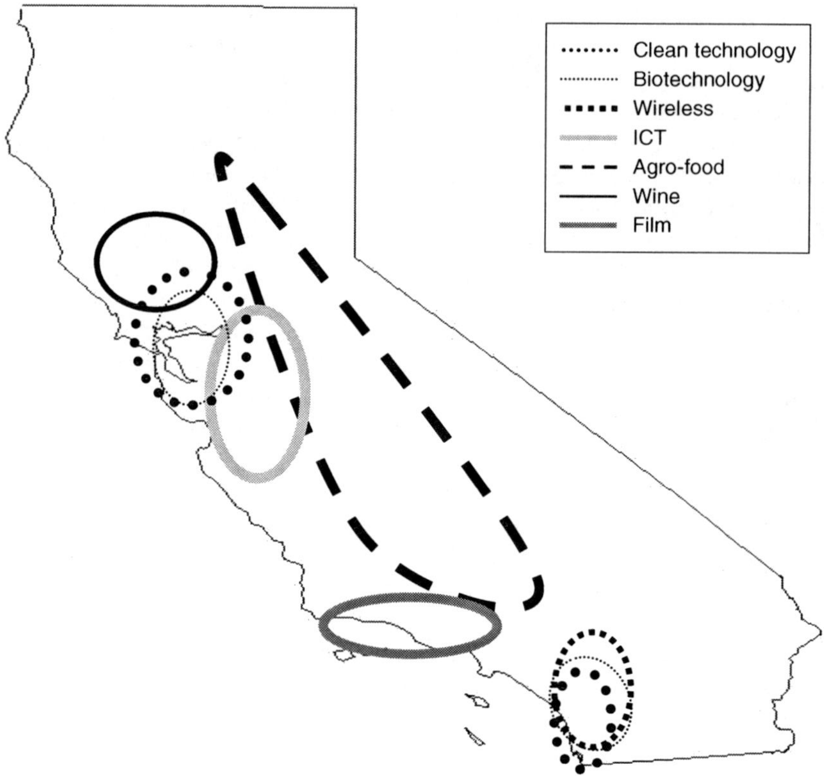

Figure 4.1 California's Jacobian clusters

recently 'discovered' cluster has all the required characteristics to warrant the cluster designation, conjoining university research at, for example, Ålborg and Århus Universities, the Danish Technological Institute (DTI) also at Århus, and both spinout firms and larger, indigenously established firms that are involved in 'green innovation'.

In Denmark the 'cluster' has no geographical specificity of the kind Porter (1998) was rather more sensitive towards. He referred there to a cluster as:

> a geographically proximate group of interconnected companies and associated institutions in a particular field, linked by commonalities and complementarities

With regard to such clusters the most important analytical task is to establish the extent of interconnections, commonalities and complementarities since this is what distinguishes a localized cluster, its specialization or differentiation and its potential for exploiting knowledge spillovers for competitive advantage. In the research to be reported below, the Danish Wind Energy Association database was accessed and the details mapped by location and categorized according to point in the supply chain. Thus final assemblers were differentiated from major module suppliers (e.g. fibre-glass blade manufacturers) and general components suppliers from them and services and logistics suppliers. Some 50 of the 70 members were found to be located in geographic proximity in Jutland, mostly in the more northerly part.

Of considerable interest here is Denmark's political commitment since 1970 to wind energy and what research reveals to be the North Jutland region's wind-turbine cluster has to be addressed. On this, Andersen et al. (2006) point to the wind energy industry having passed through an early phase characterized by numerous small and medium-sized enterprises (SMEs) producing domestically-scaled wind power for individual farms and householders. But latterly, especially since the government subsidy to domestic consumers was removed in 2000, exports have risen, the scale of equipment has increased tenfold and sea power from large-scale offshore wind farms has come to predominate. As wind turbines have only some ten years' life expectancy, most early wind turbines in rural Denmark will soon disappear if they have not already done so. So the current industry structure is large Danish (Vestas) or foreign (Siemens, Gamesa, Suzlon) producers and a supply platform of SMEs. There may be less local sourcing of key equipment like gearboxes than in the early days when North Jutland shipbuilding firms could adapt to meet the nascent wind energy demand. However, the scale and adaptability of German heavy engineering in cranes and related equipment means they now supply the Danish wind energy input market. Services and special logistics firms, the latter capable of transporting the now typically massive fibreglass turbine blades also exist in proximity as do a great many components suppliers (Fig. 4.2).

Stoerring (2007) agrees with this evolutionary profile pointing out that scale was also partly induced in the early 1980s by huge demand for wind turbines from the USA and more particularly California. Then, in the late 1980s this market collapsed because California's state administration removed its subsidy regime and the Reagan administration cut alternative energy research budgets. At this time many US turbines malfunctioned badly and even the superior Danish three-blade design was prone to breakdowns. Thereafter, the industry recovered as demand in European and Asian markets rose. Nowadays (Figure 4.2) around half global production capacity is accounted for by Danish firms like world leader Vestas

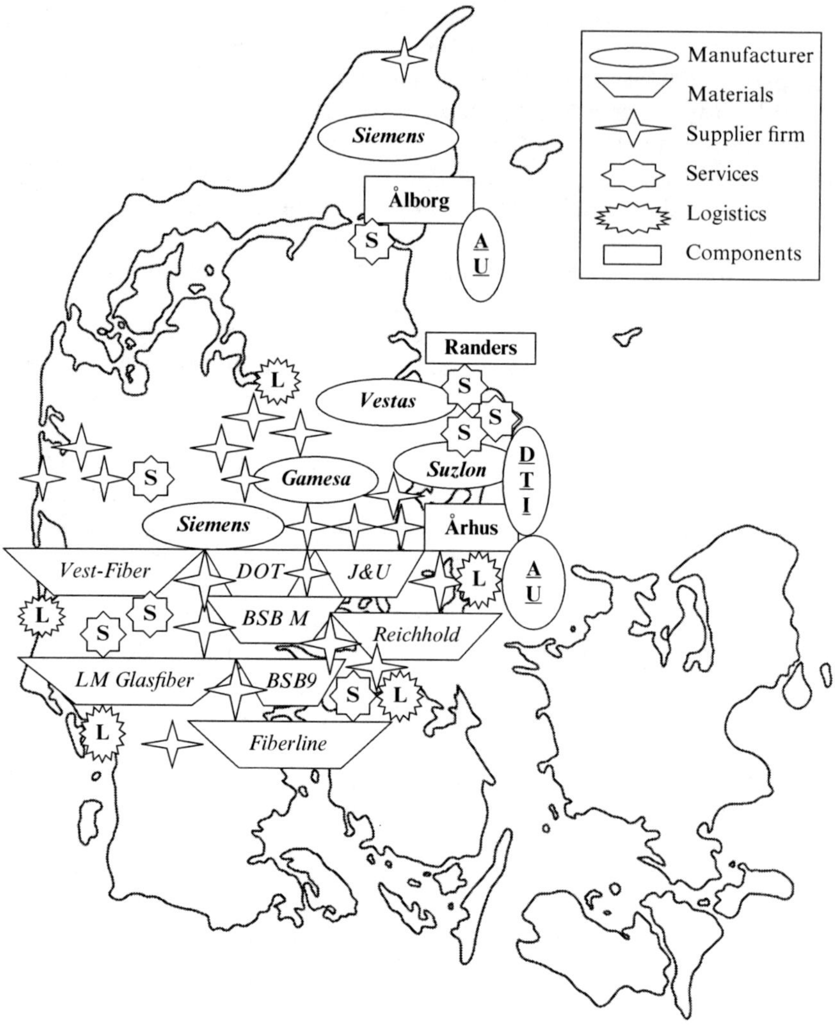

Source: Danish Wind Industry Association statistics.

Figure 4.2 The north central Jutland wind turbine cluster

Wind Systems of Randers, near Århus (acquirer of Danish firms NEG-Micon; Nordtank; Wind World) and Siemens (Bonus) at Brande and Ålborg. Gamesa Wind Engineering, Spain's largest producer of turbines is at Silkeborg. Jutland. Suzlon, India's leader is located at Århus. LM Glasfiber of Lunderskov near Århus in Jutland is the leading supplier of

Source: Composed from ESTIF data by the author.

Figure 4.3 North Jutland's solar thermal energy cluster

fibreglass wind turbine blades. The other members of the North Jutland cluster are in Figure 4.2. Of the Danish Wind Industry Association's 70 members, 50 are in Jutland, mostly north-central Jutland. More is said on the etymology of this 'green cluster' evolution in the final section of this chapter. Universities (*AU*) join DTI (see below) as a knowledge generation subsystem of the RIS.

As noted earlier, overlapping this substantial and globally leading wind turbine technology cluster is the main Danish solar thermal energy cluster (Figure 4.3). This is smaller in scope but consists of largely indigenous firms and their suppliers. These involve firms in two types of supply chain as follows:

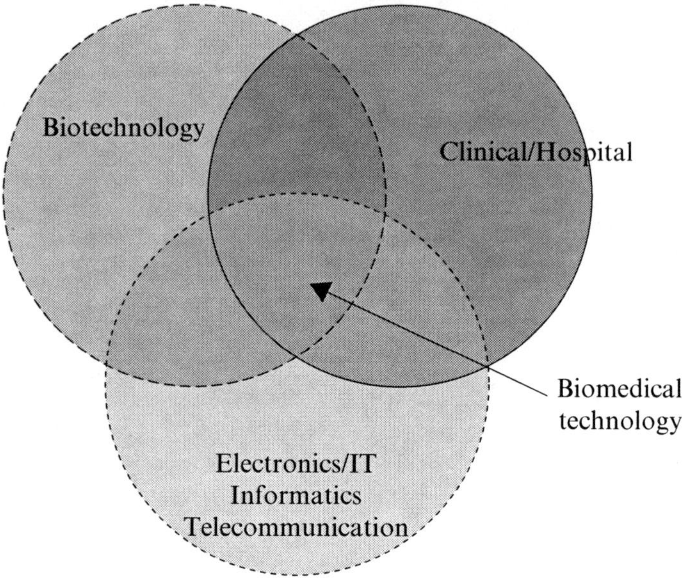

Figure 4.4 Jacobian cluster emergence in North Jutland

Solar collectors

- Glazed (roofs)
  - Flat plate collectors
    - Glass
    - Heat absorbent copper/aluminium
    - Coatings, paint
    - Pipes welded to absorber plate
  - Vacuum collectors
    - Parallel glass tubes
    - Absorber
    - Transfer pipes
    - Vacuum is insulator
- Unglazed (swimming pools) long tubes
    - Synthetic absorbent material
    - Hydraulics in pool filtration system
- Heat storage and back-up heating
- Plumbing and installation

Finally, exemplifying North Jutland's Jacobian cluster profile it is worth considering Figures 4.4 and 4.5, the first of which reveals established cluster evolution in the shape of the NorCom wireless communications cluster at

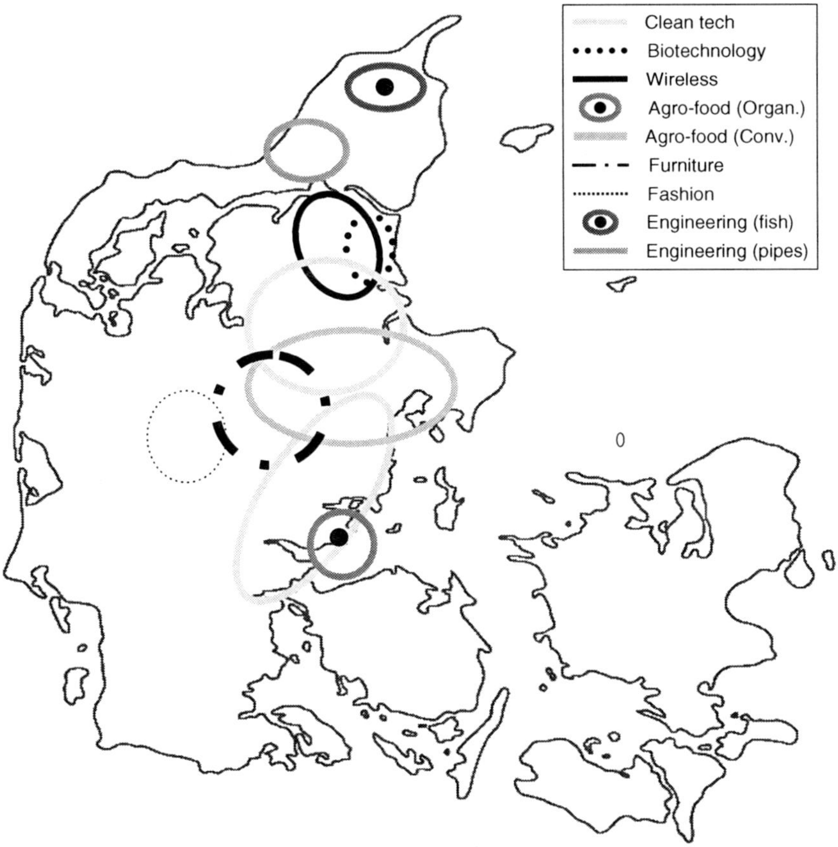

Figure 4.5 North Jutland's Jacobian clusters and related variety

Ålborg and the possibly emergent and overlapping biomedical technology cluster in close proximity (Stoerring, 2007). Here, the long-established wireless telecommunications cluster (Stoerring and Dalum, 2007) has given rise to possible cluster mutation by interaction with the healthcare activities associated with clinical trials and testing of biomedical equipment. Many of these activities are closely associated with science and technology commercialization through academic entrepreneurship at Ålborg University. In Figure 4.5 are shown the most prominent (though many have yet to be fully researched) of North Jutland's Jacobian clusters which are characterized as emergent clusters or established ones by their 'related variety' characteristics in relation to each other. This may be understood as follows in Table 4.1. In this may be seen the stylized history of a significant

Table 4.1 Jacobian cluster path dependences

Technology	Path dependence
Clean technology	Agricultural and marine engineering (e.g. wind turbine blades replicate plough and propeller design)
Biotechnology	Wireless ICT and medical technology
Wireless telephony	Ship-to-shore marine technology
Agro-food (conventional, intensive)	'Railroadization' of undeveloped land in Jutland
Organic food	Reaction against conventional intensive food production in Jutland (mostly dairy)
Furniture	'Railroadization', craft schools and the local forestry tradition
Fashion clothing	Women's craft schools skilling farmers' wives
Fish equipment and Pipework engineering	Fishing and marine engineering

part of the Jutland economy's evolution since it was radically transformed by nineteenth century railroadization as proposed by Schumpeter (1975). This process created certain path dependences or developmental trajectories. This kind of analysis is rather important and useful in explaining the ontology of such regional economies and their clusterization. Recall, for Schumpeter 'railroadization' was the purest, most radical kind of innovation based upon the creative destruction of a preceding state of nature (or at least non-farming economy). The massive 'entrepreneurial event' of 'railroadization' creates evolutionary trajectories that act as constrained opportunity sets for regional evolution. Activities displaying 'related variety' to the originating entrepreneurial event comprise the selected trajectories as in Table 4.1. These may foster varying intensities of innovation from disruptive (after Christensen, 1997) which cheapens (e.g. mobile telephony) an existing but specialized, uncommoditized technology (e.g. ship-to-shore radio) to incremental innovations around mobile telephony (first, second, third, etc. generation mobile telephony).

There is insufficient space to offer a satisfying explanation for the Jacobian cluster mutation process in North Jutland but Kristensen (1992) underlines 'railroadization' as a key process where Jutland as a whole was opened up on a smaller scale but with similar inspiration to that of the Frontier West in the nineteenth century USA. With this came two key movements. The first was the farmers' co-operative movement where farmers supplied their own production and household needs, including banks. The second movement was the craft schools, established in over 350 centres, followed by the still flourishing Danish Technological Institutes from 1907. Together these made a form of social or collective entrepreneurship possible. That is, infrastructure, education, technical support, finance and markets. Hence 'social capital' remains an important dimension of the SME-based collective entrepreneurship of North Jutland. It makes technological branching by means of related variety evolution possible. Finally, this is assisted by the existence of a RIS infrastructure of technological institutes, technical and craft schools and universities, which sustains entrepreneurship and localized knowledge transfer.

BIOENERGY FROM CROPS IN WALES

One of the most surprising, perhaps, but unquestionably innovative developments in the bioenergy field has occurred in recent years in Wales. Descriptively speaking it involves patented knowledge derived by the Institute for Grassland and Environmental Research (IGER) based at Aberystwyth in rural, central Wales. This UK Biological Research Council-funded research institute has, for 70 years to 2007 been the UK's main, specialist grassland research institute. It was tasked from the outset with improving the quality of fodder for cattle and sheep feedstock, which is mainly grass. By the early 1980s, research, which involved not simply breeding richer grasses but understanding the rumen of these ruminative animals, had revealed that a limit to quality on these mountain-bred animals occurred because the enzymes that broke down fodder into protein were actually consuming a significant portion of the nutritional value of the fodder consumed by the animal. Following many years of lengthy field trials and laboratory research, cross-breeding the basic rye-grass commonly utilized for cattle and sheep fodder with breeds possessing enhanced sugar content produced optimal results. The enzymes took some of the enhanced sugar content, transforming it directly into energy but left a substantial portion for the animal sufficient for the amount, nutritional value and flavour of the animal to be significantly enhanced. This came to the market at a time when consumer demand for leaner meat of the type raised in mountainous areas rose significantly and continuous improvement to the original AberDart strain of rye grass, marketed by Germinal Holdings, over the intervening years led to it reaching 50 per cent of the UK market. It further secured the status of Welsh Black beef and Welsh lamb as premium products and enabled significant improvements to occur in comparable upland cattle breeds such as Aberdeen Angus.

In 2003, it was realized that IGER had, in the form of these SugarGrasses, an indigenous product to add to its burgeoning portfolio of biofuels. Tests had shown that SugarGrass had twice the calorific value of sugar cane, the source of much of the world's biofuel. IGER thus evolved a second string to its grassland expertise by developing a renewables research division. One of the biofuel feedstocks in which it became supreme early on was the growing and processing of *Miscanthus*, more popularly known as Elephant Grass, an African tall grass that grows on marginal land. Accordingly it doesn't compete for land with food crops, one of the criticisms of the USA and Europe's 'bolt for biofuels'. This has seen the ears and cobs of wheat and corn being turned into ethanol because of easy availability and major subsidy, causing up to 40 per cent increases in the price of such cereals, and grief in developing country food markets.

Tellingly, IGER is widely perceived as in a global class of its own in these specific bioenergy subfields, the official view being that maybe University of California, Berkeley, may become competitive now they have received a $500 million endowment for a Climate Change research institute from British Petroleum (BP). Apart from the University of Illinois, also mentioned as a possible future competitor, but only those two – IGER has a current lead on both of them. But in any case, SugarGrass is also twice as calorific as *Miscanthus* and SugarGrass is thus favoured as the technology with the best long-term prospect to replace oil. And IGER has the patent for SugarGrass, currently earning royalties of £100 000 per year from sales of seed varieties for fodder. But as the world awakens to the relatively simple processes of biorefining the product, these are likely to grow substantially.

So much so that agreement has been reached with Welsh government officials about the promise of funds to help build an experimental biorefinery. Thinking had gone as far as to speculate that when oil ceases to be refined at the huge Milford Haven refineries in neighbouring Pembrokeshire, the pool of talent and infrastructural sunk costs would make them ideal candidates for becoming SugarGrass (and *Miscanthus*) biorefineries. These would continue to meet a huge share of the UK's future energy. But it is not simply a spinout–venture capital model that is in mind. Possibly because a spinout model doesn't yet work as well as a commercialization outsourcing model in this nascent field. For example, *Molecular Nature*, the key spinout of IGER, burnt up its venture capital. But because of the value of its patent for biofuels potential as well as its fodder market, it was acquired by spin-in company *Summit*. Moreover, true to the traditions of co-operation among Welsh mountain farmers,

Figure 4.6 Solar energy equipment manufacturers in Wales

IGER promotes a new vision of mixed farming whereby groups of farmers grow *Miscanthus* on their poorest soil, devote some fields for SugarGrass fuel cropping and raise quality Welsh Lamb or Welsh Black Beef on their best SugarGrass land.

Photovoltaics produce solar thermal energy as in North Jutland. In Wales, this has been studied by authors (Hendry et al., 2001) comparing the broader opto-electronics cluster, which also specializes in fibre-optic cabling, with those such as that associated with Carl Zeiss in Jena, eastern Germany. However, in relation to this present discussion about 'green clusters', it is the photovoltaics capability that comes to the fore. Figure 4.6 reveals the presence of subdivisions of multinationals such as Japanese electronics corporation Sharp whose Sharp Solar subsidiary is based at St Asaph alongside Corus Colours, a subsidiary of Corus, the UK–Dutch steel manufacturer, acquired in 2007 by Indian giant Tata Steel. Utilizing polymer science and surface treatments Corus Colours has innovated radically a Solar Paint product capable of generating solar energy, especially from prefabricated steel buildings. Other firms in the photovoltaics cluster at St Asaph are indigenous, such as Cardiff-headquartered microprocessor firm IQE and 'green engineering' firm Dulas, headquartered in mid-Wales.

Hence, in conclusion, we see that numerous indications of clustering among small firms, but also some large firms, along with an applied and basic research infrastructure characterises important locations of 'green clusters' mainly, in this analysis, focused upon the production of non-fossil fuel energy that contributes to the moderation of global warming. A key feature to be discussed in the concluding section of this chapter is that in some cases there is an element of cluster 'species' multiplication which, from an evolutionary economic geography perspective can readily be hypothesized. As shown in Figure 4.1, the California 'Cleantech' clusters are to be found in juxtaposition to the ICT and biotechnology, food and wine clusters of the San Francisco region of northern California and the wireless telecom and biotechnology clusters of San Diego in southern California. Indeed, so-called Cleantech is widely seen as arising from the combination of biotechnology (including biopolymers and biofuels), ICT (sensors) and nanotechnology (catalysts and filtration membranes). However, while agro-food is also one of California's key industries, agro-food path dependence seems even more pronounced in the cases of Jutland and Wales, as we have seen, while in yet another case, forestry is important to Sweden's biofuels cluster in Örnsköldsvik (Cooke, 2007).

CONCLUSIONS AND THEORETICAL IMPLICATIONS

By virtue of an examination of the emergence of green clusters, often involving the production of new forms of non-fossil fuel energy aimed at lessening of overall GHG emissions derived from human economic activity, a curious feature of economic evolution has been revealed. The clue lies in the element of convergence that characterises green innovation. As hinted at in the cases of northern and southern California, not studied in detail here but examined elsewhere (Cooke, 2007), the type of Cleantech industry emerging in the clustered form described by Burtis et al. (2004, 2006) evolves from agro-food, ICT and biotechnology. In North Jutland we see something comparable having occurred. Thus the wind turbine and solar thermal clusters are found in the more agricultural and marine engineering regions of Denmark. In writing the history of the former industry, Karnøe and Jørgensen (1996) and Jørgensen and Karnøe (1995) note how the Danish design of wind turbines defeated the main global competitor from where a significant renewable technology demand also arose simultaneously from the 1970s, namely California. As noted, Danish wind turbine blade design was influenced by the agricultural engineering industry, notably the design of modern ploughing equipment. In the experimental innovation phase when some 30 firms engaged in the design of prototype turbine blades, knowledge spillovers from the design of propellers by marine engineers in the Jutland shipbuilding industry were also absorbed. This resulted in a three-blade solution and the idea that the greater efficiency in the operation of such blades came from pointing them into the wind. California's aeronautics tradition, by the 1970s predominantly relying on jet propulsion, led to the recovery of historic knowledge of propeller-driven aeroplanes. This suggested a two-blade solution pointing downwind. The Danish solution proved far superior to the Californian in this technological contest.

Hence in these multi-cluster locations, it is clear that a good deal of technological convergence is possible and probably necessary. But, interestingly, comparable technological assets do not necessarily produce optimum solutions from such Schumpeterian 'new combinations'. Nevertheless, it is clear that in some regions, cluster forms can evolve quite readily from other cluster forms, the cluster 'species' multiplication giving the region more of a cluster 'platform' characteristic to its industrial organization. On further inspection, both California and Jutland prove to have spawned many clusters. In the former case, wine clusters overlap the horticultural zones, Hollywood's film cluster is well-known and Porter (1998) also profiles other, sometimes highly specialized clusters such as the alloy golf club cluster at Carlsbad in the southern Californian desert. Further inspection of the cluster history of Jutland reveals the detailed cases of Salling (furniture) and Ikast (clothing), the even more closely studied NorCom wireless telephony cluster at Ålborg (Stoerring and Dalum, 2007), the emergent biomedico cluster also at Ålborg, and as yet unexamined cluster candidates in insulated pipework near Ålborg, and fish processing equipment near Skagen, at Jutland's northern tip. At Barritskov, east Jutland is the estate that sustains the Årstiderne Organic Food Network, a co-operative retail network that delivers 30 000 boxes per week of organic food throughout Denmark. It could also be argued that there is a high degree of knowledge transfer from varieties of agricultural production to bioenergy production in Wales leading to possibly nascent cluster-formation, but also from glass technology to fibre-optic cables and then photovoltaics by a different route into renewable energy in a multifunctional opto-electronics cluster. Species multiplication or mutation of this kind would be perfectly consistent with an underlying theory of evolutionary economic geography, especially that part referring to the opportunities for innovation and growth arising where there is *related variety* among industries. Absorptive capacity for adaptation to new combinations based on easily understood knowledge spillovers would be the mechanism by which such species multiplication is explained, as the case of Jutland's wind turbine technology illustrates especially clearly.

In other cases, focusing upon 'green innovation' cluster specialization rather as ascribed to Marshall–Arrow–Romer (MAR) thinking seems on the face of it to be more convincing than the idea of Jacobian clustering (after Jane Jacobs' notion of innovation through variety). Yet even where limited clustering occurs, as in Rhineland or Brazil previously existing industries, whether the coal, steel and chemicals super-clusters of the Ruhr Valley or the sugar producing industry in Brazil are suggestive of the presence of important spillovers from knowledge of filtration and ventilation in the former and fermentation in the latter cases that were of profound importance to the evolution of new, convergent combinations of innovative products and processes. This tends to confirm clearly the widespread and common-sense policy experience that clusters cannot be easily built *in vacuo* but may find it a less rigorous evolutionary trajectory to emergence where the regional context gives opportunities for Schumpeterian 'new combinations' from regionalised 'related variety'. Where such related variety is more attenuated, as perhaps with biofuels in Brazil or NE England, fewer 'Jacobian clusters' emerge.

However, that is not the whole of the explanation for Jacobian cluster mutation, rather it is an important contextual factor as noted, for example in the work of Cantwell and Iammarino (2003). Other key features that may be hypothesised, but further research is needed, is that Jacobian clustering benefits from other more social, institutional, and organizational assets, such as those listed below, in addition to more economic assets concerning related variety, knowledge spillovers and high lateral absorptive capacity:

- Social capital
- Collective entrepreneurship
- Technological branching ('new combinations' opportunities)
- Peripherality (perceived distance from key governance core)
- Infant industry subsidy
- Innovation system – regional research and technological institutes, universities, regional innovation platform policy and funding.

The key concluding point of this section is that, for the first time in regard to new industries, we see replication of processes that have historically underpinned successful regional economies that once spawned many traditional industrial districts or clusters. Evolutionists like Klepper (2002) for example would also highlight the transfer of routines from one to another industry by means of 'mobility of talent', as in the cases of the US, German and Italian automotive and engineering industries (see also Boschma and Wenting, 2007). Probably the key findings of this contribution in relation to evolutionary theory are the following. First, while Schumpeter had little to say about regional innovation, his concept of innovation by 'railroadization' proves to be highly apposite as an explanation of at least the case of Denmark's opening up of North Jutland and elsewhere in the west in the nineteenth century and its modern evolution into an arena of Jacobian clustering in related variety industries. Second, the green perspective somehow threw the evolution of this kind of industry organization into clearer perspective because it focuses on a horizontal and convergent technology 'platform concept' rather than a more traditional industrial economics perspective that emphasizes vertical structures like sectors or clusters. Finally, regarding cluster emergence within a regional innovation systems context, the research reported showed the importance of social capital, which even in California may be considered strong, as the work of Saxenian (1994) on Silicon Valley showed, as an evolutionary driver of certain kinds of regional innovation system. Indeed, whether as 'bonding', or more institutional 'bridging' social capital it is the key element of the hidden power of networks, both social and institutional, that has always been at the heart of the RIS approach to evolutionary science. Finally, it could be seen that the evolutionary processes described were capable of hosting differing intensities of 'innovative bursts'. Railroadization itself was said by Schumpeter to be the most radical kind. Divergent, possibly disruptive, innovations like the semiconductor in California, mobile telephony infrastructure in N. Jutland cheapened and 'democratized' key technologies based upon new knowledge combinations. Thereafter, incremental, narrowly path-dependent innovation can evolve, or there may be punctuated evolution with the more radical innovations around biotechnology from cancer-defeating therapeutics to fodder-based biofuels as knowledge evolves and broader economy regimes, notably that associated with the chemicalization of fossil fuels, approach exhaustion and make way for a potentially cleaner bioeconomy regime.

NOTE

1. The Myrdal–Hirschman theory of economic development has been influential in the emergence of 'new economic geography' (e.g. Krugman, 1995). Anticipating the latter's

solution to the neoclassical location theory impasse by positing 'increasing returns to scale' rather than the rubric of 'constant returns' thus demonstrating the growth of cities to be a function of spatial monopoly, Myrdal (1957) proposed spatial development to be characterized by 'cumulative causation' with associated 'spread' and 'backwash' effects. This implies increasing returns to scale (through 'backwash') and developmental 'spread' to other nearby areas. Hirschman's (1958) elaboration on this was that 'spread' would be driven by the innovative capacity of competing technology users. Under 'knowledge economy' conditions we hypothesize that, over relatively short time periods, primate cities grow through increasing returns (to knowledge) and 'satellites' of leading technology innovators 'spread' nearby. Our preliminary static pictures of EU NUTS 2 regions (Cooke and Schwartz, 2008) are consistent with this, while our dynamic picture of spatial divergence in Israel 1995–2002 is consistent with Myrdal–Hirschman rather than Krugman (2001), who himself admits his 'two-locations competing' models are misleadingly simplistic. In this respect, it can be argued, evolutionary economic geography trumps 'new economic geography'.

REFERENCES

Andersen, E. (1994), *Evolutionary Economics: Post-Schumpeterian Contributions*, London: Pinter.

Andersen, E. (2007), *Schumpeter's Evolution*, Ålborg: Ålborg University.

Andersen, P., M. Borup and M. Olesen (2006), 'Innovation in energy technologies', *Risø Energy Report*, **5**, 21–7.

Antonelli, C. and F. Momigliano (1981), 'Problems and experiences of regional innovation policy in Italy', *Micros*, **2**, 45–58.

Bergman, E., G. Maier and F. Tödtling (eds) (1991), *Regions Reconsidered: Economic Networks, Innovation and Local Development in Industrialized Countries*, London: Mansell.

Boschma, R. (2005), 'Proximity and innovation: a critical assessment', *Regional Studies*, **39**, 61–74.

Boschma, R. and R. Wenting (2007), 'The spatial evolution of the British automobile industry: does location matter?', *Industrial and Corporate Change*, **16**, 213–38.

Burtis, P., R. Epstein and R. Hwang (2004), *Creating the California Cleantech Cluster*, San Francisco, CA: Natural Resources Defence Association.

Burtis, P., R. Epstein and N. Parker (2006), *Creating Cleantech Clusters*, San Francisco, CA: Natural Resources Defence Association.

Cantwell, J. and S. Iammarino (2003), *Multinational Corporations & European Regional Systems of Innovation*, London: Routledge.

Castells, M. (1996), *The Rise of the Network Society*, Oxford: Blackwell.

Commission of the European Communities (CEC) (2001–07), *Regions: Statistical Yearbook 2001*, Luxemburg: Commission of the European Communities.

Chesbrough, H. (2003), *Open Innovation*, Boston, MA: Harvard Business School Press.

Christensen, C. (1997), *The Innovator's Dilemma*, Boston, MA: Harvard Business School Books.

Cooke, P. (1985), 'Regional innovation policy: problems and strategies in Britain and France', *Environment and Planning C: Government and Policy*, **3**, 253–67.

Cooke, P. (1992), 'Regional innovation systems: competitive regulation in the new Europe', *Geoforum*, **23**, 365–82.

Cooke, P. (1993), 'Regional innovation systems: an evaluation of six European cases', in P. Getimis and G. Kafkalas (eds), *Urban and Regional Development in the New Europe*, Athens: Topos.

Cooke, P. (2002), *Knowledge Economies*, London: Routledge.

Cooke, P. (2004), 'Introduction: regional innovation systems – an evolutionary approach', in P. Cooke, M. Heidenreich and H. Braczyk (eds), *Regional Innovation Systems*, London: Routledge.

Cooke, P. (2007), *Growth Cultures: the Global Bioeconomy and its Bioregions*, London: Routledge.

Cooke, P. (2008a), 'Green clusters: green innovation and Jacobian cluster mutation', *Cambridge Journal of Regions, Economy & Society*, 1 (forthcoming).

Cooke, P. (2008b), '*Cleantech* and an analysis of the platform nature of life sciences: further reflections upon platform policies', *European Planning Studies*, 16, 3 (forthcoming).

Cooke, P. and A. da Rosa Pires (1985), 'Productive decentralisation in three European regions', *Environment and Planning A*, **17**, 527–54.

Cooke, P. and D. Schwartz (2008), 'Regional knowledge economies: an EU–UK and Israel perspective', *Tijdschrift Voor Ekonomische Geographie*, **99**, 178–92.

Cooke, P., R. Alaez and G. Etxebarria (1991), 'Regional technological centres in the Basque country: an evaluation of policies, providers and user perceptions', *Regional Industrial Research Report No. 9*, Cardiff: Cardiff University.

Dosi, G., C. Freeman, R. Nelson, G. Silverberg and L. Soete (eds) (1988), *Technical Change and Economic Theory*, London: Pinter.

Grabher, G. (1991), 'Building cathedrals in the desert: new patterns of cooperation between large and small firms in the coal, iron and steel complex of the German Ruhr area', in E. Bergman, G. Maier and F. Tödtling (eds), *Regions Reconsidered: Economic Networks, Innovation and Local Development in Industrialized Countries*, London: Mansell.

Guthman, J. (2005), *Agrarian Dreams: the Paradox of Organic Farming in California*, Berkeley, CA: University of California Press.

Hendry, C., J. Brown, H.-D. Ganter and S. Hilland (2001), 'Industry clusters as a location for technology transfer and innovation: the case of Opto-Electronics', *Industry and Higher Education*, **15**, 33–41.

Hirschman, A. (1958), *The Strategy of Economic Development*, New Haven, CT: Yale University Press.

Jacobs, J. (1969), *The Economy of Cities*, New York: Vintage.

Johansson, B. (1991), 'Economic networks and self-organization', in E. Bergman, G. Maier and F. Tödtling (eds), *Regions Reconsidered: Economic Networks, Innovation and Local Development in Industrialized Countries*, London: Mansell.

Jørgensen, U. and P. Karnøe (1995), 'The Danish wind turbine story: technical solutions to political visions', in A. Rip, T. Misa and J. Schot (eds), *Managing Technology in Society: The Approach of Constructive Technology Management*, London: Pinter.

Karnøe, P. and U. Jorgensen (1996), *The International Position and Development of the Danish Wind Turbine Industry*, Copenhagen: AKF.

Klepper, S. (2002), 'Capabilities of new firms and the evolution of the US automobile industry', *Industrial and Corporate Change*, **11**, 645–66.

Kristensen, P. (1992), 'Industrial districts in West Jutland, Denmark', in F. Pyke and W. Sengenberger (eds), *Industrial Districts and Local Economic Development*, Geneva, Switzerland: International Institute for Labour Studies.

Krugman, P. (1995), *Development, Geography and Economic Theory*, Cambridge, MA: MIT Press.

Krugman, P. (2001), 'Where in the world is the "new economic geography"?', in G. Clark, M. Feldman and M. Gertler (eds), *The Oxford Handbook of Economic Geography*, Oxford: Oxford University Press.

Lundvall, B. (1988), 'Innovation as an interactive process', in G. Dosi, C. Freeman, R. Nelson, G. Silverberg and L. Soete (eds), *Technical Change and Economic Theory*, London: Pinter.

Malecki, E. (1991), *Technology and Economic Development*, London: Longman.

Meyer-Krahmer, F. (1990), *Science and Technology in the Federal Republic of Germany*, London: Longman.

Myrdal, G. (1957), *Economic Theory and Underdeveloped Regions*, London: Duckworth.

Porter, M. (1990), *The Competitive Advantage of Nations*, New York: Free Press.

Porter, M. (1998), *On Competition*, Boston, MA: Harvard Business School Press.

Rothwell, R. and M. Dodgson (1991), 'Regional technology policies: the development of regional technology transfer infrastructures', in J. Brotchie (ed.), *Cities of the 21st Century*, London: Longman.

Saxenian, A. (1994), *Regional Advantage*, Cambridge, MA: Harvard University Press.

Schumpeter, J. (1975), *Capitalism, Socialism and Democracy*, New York: Harper.

Scott, A. (2006), 'Spatial and organizational patterns of labour markets in industrial clusters: the case of Hollywood', in B. Asheim, P. Cooke and R. Martin (eds), *Clusters and Regional Development: Critical Reflections and Explorations*, London: Routledge.

Simard, C. and J. West (2003), 'The role of founder ties in the formation of San Diego's "Wireless Valley"', paper to DRUID Summer Conference 2003; Creating, Sharing and Transferring Knowledge: the Role of Geography, Organizations and Institutions, Copenhagen, 12-14 June.

Smith, K. (2000), *What Is the Knowledge Economy?*, Brussels: European Commission.

Stoerring, D. (2007), 'Emergence and growth of high technology clusters', PhD thesis, Department of Business Studies, Aalborg University.

Stoerring, D. and B. Dalum (2007), 'Cluster emergence: a comparative study of two cases in North Jutland, Denmark', in P. Cooke and D. Schwartz (eds), *Creative Regions: Technology, Culture and Knowledge Entrepreneurship*, London: Routledge.

5. Higher education and economic development: do we face an intertemporal trade-off?

James R. Wilson

1. INTRODUCTION

The production of research and learning activities that takes place in universities represents a strategically important area of economic activity; in most economies, higher education (HE) fulfils various key roles in advancing economic development. Most widely recognised are the contributions of the HE sector in providing economies with certain forms of human capital and in uncovering potentially significant research advances. In turn both of these feed into the development of other economic sectors and of society in general. Thus universities are seen as vital components in regional and national systems of innovation, for example. In many countries HE is also increasingly recognised as having significant potential as a generator of economic rents (and export revenues) in its own right. Finally, universities are sometimes acknowledged for their contribution to the development of characteristics such as capacity for critical thought and open-mindedness among the people of a society.

Recent years, however, have seen quite dramatic changes in the organisation of the HE sector in many parts of the world; in particular it is argued that a 'commercialisation' of universities has been (and is) taking place (Bok, 2003). In this chapter we reflect in particular on experience from the UK in analysing these changes. Given the whole range of potential contributions of HE to an economy, we explore the potential implications of these changes for economic development.

We build from the strategic choice approach to economic development (Cowling and Sugden 1998, 1999; Sugden and Wilson, 2002, 2005; Bailey et al., 2006) to analyse the roles that universities might be expected to play in an economy. It is argued that in addition to their more traditionally acknowledged contributions to economic development, the strategic choice approach highlights the significance of another role: universities can contribute to the capacity of a society to engage in critical thought and deliberation around key decisions impacting on development processes. From this theoretical context, the chapter explores implications of fundamental and significant developments in the sector that are emerging particularly rapidly. It is argued that the commercialisation of HE can be untangled into four interrelated trends: 'marketisation', 'globalisation', 'corporatisation' and 'privatisation'. The chapter identifies and analyses each of these, and we discuss implications for the different roles that the HE sector might desirably play within an economy. In particular, it is suggested that aspects of these processes affect the capacity of universities to stimulate independent, critical and creative thought among faculty and students, and therefore within society more generally. Relating back to strategic choice arguments, we ask whether recent developments therefore impact on the potential contribution of universities to economic development processes.

Section 2 begins with a brief exposition of the strategic choice approach to economic development. Section 3 then outlines the potential contributions of the HE sector to an economies' development, introducing the argument that universities can contribute to societal capacity for engagement in critical thought and deliberation, enhancing the democratic potential in strategic choice processes. Section 4 focuses on current trends in universities, and suggests an untangling of the concept of 'commercialisation'. The identification and analysis of these trends reflects in particular on UK experience. In section 5 the chapter concludes with a discussion of the implications of these identified trends for the roles that HE is able to play in an economy. In particular we argue the possibility that, in the UK at least, the sector is facing an inter-temporal trade-off between short-term market 'competitiveness' and longer term ability to effectively fulfil more fundamental developmental roles in the economy.

2. STRATEGIC CHOICE AND ECONOMIC DEVELOPMENT

The premise of the strategic choice framework in analysing economic development is that concentration of decision-making power in economies implies likely failure to meet the development aims and objectives of societies. This is rooted in Cowling and Sugden's (1987, 1998, 1999) analysis of the firm, and in particular the transnational firm. They draw on Coase's (1934) seminal work on the theory of the firm and Zeitlin's (1974) observations on corporate control to argue that the modern corporation can be conceptualised as a 'centre of strategic decision making'. Moreover, given a reality of imperfect markets,[1] the concentrations of power inherent in these centres of decision-making imply 'strategic failure' in economies. As summarised by Bailey et al. (2006, p. 559): 'the concentration of strategic decision-making power implies a failure to govern production in the public interest, defined as the agreed upon, evolving concerns amongst all of those indirectly and significantly affected by the production activities . . . . The idea is that strategy in practice is decided upon by exclusive interests who ignore the concerns of those indirectly and significantly affected by their decisions'.

The strategic failure concept has been applied to a range of specific sectors and circumstances (Sacchetti and Sugden, 2003; Branston et al., 2006b, 2006c; Branston and Wilson, 2006; Cowling et al., 2009), including to the internationalisation of universities (Sugden, 2004). More generally, Sugden and Wilson have critiqued contemporary economic development policy (2002) and globalisation (2005) from this perspective. For example, it is argued that current globalisation can be characterised as an elite form, in which outcomes are determined primarily by the exclusive interests that govern powerful firms and institutions in the global economy. Potential alternative forms of globalisation may provide opportunities for wider groups of people to influence their development outcomes. Essentially there are choices to be made, or to follow Coe and Yeung (2001, p. 368), current economic globalisation is not an 'immutable inevitability, but a set of processes that is socially constructed, and therefore can be encouraged or resisted by actors/institutions at various scales'. Indeed, the principle of voluntarism in economic development is a fundamental assumption of the strategic choice framework, which is 'predicated on there being actual and potential scope both for people to influence the set of choices available, and for them to exercise choice' (Bailey et al., 2006, p. 557). Variation in the scope for choice is therefore a distinguishing feature between economic realities; for example between different forms of globalisation, or more specifically between different firm, industry or local realities.

In terms of policy significance, a central argument emerging from this literature is that the avoidance of strategic failure necessitates the widening of opportunities and capacities for people to participate in decision-making processes that effect their development. Put another way, if development processes are to be socially efficient, in the sense of outcomes reflecting the aims and objectives of people, then attention must be paid to how choices are made in economies and their constituent industries, firms and institutions. In particular, there is a role for policy in enhancing the capacity among people to inform themselves of the issues around key decisions, and to engage with one-another in influencing the decisions that determine development paths. Following Hirschman (1970) this might be thought of in terms of a move from over-reliance on the mechanism of 'exit' in decision-making to facilitating the effective development of 'voice'. It is recognised, however, that this is not a simple task in a predominantly market economy; Hirschman (1970) for example describes 'voice' as 'essentially an *art* evolving in new directions' (p. 43), and suggests that the contrasting simplicity of 'exit' creates a bias that tends to 'atrophy the development of the art of voice' (p. 47, emphasis removed).[2]

Branston et al. (2006a) and Sacchetti and Sugden (2007) draw on the work of Dewey (1927), and on Long's (1990) analysis of his work, in exploring these issues. They argue that strategic decisions have consequences for the 'private' interests of the agents directly undertaking an activity, and potentially also for various affected 'public' interests. For Branston et al. (2006: 195), public interests in the activities of a firm are 'the agreed upon, evolving concerns among all of those indirectly and significantly affected by those activities'. Habermas's (1964: 116) influential notion of the 'public sphere' as 'the realm of our social life in which something approaching public opinion can be formed' is also interesting. Curran (1991, p. 29) interprets this as 'the space between government and society in which private individuals exercise formal and informal control over the state: formal control through the election of governments and informal control through the pressure of public opinion'.

Whether we are considering the relationship between firms and 'publics' or governments and 'publics', the fundamental problem, as identified by Dewey (1927), is similar: *how* do 'publics' identify and organise themselves so as to determine and express their 'interest' in decision-making processes? In this regard there is an important strand of literature in democratic theory stressing the significance of deliberative forms of democracy (Elster, 1998; Bohman, 1998; Warren, 2005). As highlighted by Warren (2005, p. 194), deliberation 'reveals the private information necessary to make collective decisions'. Thus it is a process that complements the power to make decisions inherent in mechanisms such as voting and markets. It is beyond the scope of this chapter to provide a detailed discussion of literature on deliberative democracy, but its analysis clearly has relevance for the development of processes within economies that are capable of identifying and expressing 'public interests'. In particular, while current research is unclear on exactly what is required, it is evident that long-term and embedded processes of learning are critical foundations (Wilson, 2004; Sugden and Wilson, 2005). Alongside the widely accepted significance of learning for bridging knowledge gaps and generating competitive advantage, this analysis emphasises learning in an active, dynamic and creative sense; learning that is capable of contributing to the challenges of identifying and expressing public interests.

It is from this departure point that we suggest higher education has a potentially important role to play. Similarly to the media industry (on which see Branston and Wilson, 2006; also see Chapter 11 of this volume), it is reasonable to assert that the HE sector can have a potentially sizeable impact on the capacity to generate and make use of opportunities for publics to identify themselves, inform themselves, and exercise their influence on key socio-economic decisions. Such arguments do not typically feature explicitly in analysis of the contributions of the higher education sector to the economy. We turn now in section 3 to analyse the conventional arguments, before introducing an additional important contribution that follows from this strategic choice perspective and that we suggest is in danger of being overlooked in the context of current industry trends.

3. CONTRIBUTIONS OF THE HE SECTOR TO THE DEVELOPMENT OF ECONOMIES

Higher education constitutes a key economic sector in most economies, although the balance of its contributions varies from economy to economy. While its production of tangible goods and services is not as obvious as many other sectors, there are well-established reasons for its strategic importance in contributing to economic development. First, it plays a crucial role in facilitating the development of knowledge and skills among the people of an economy (human capital development). In turn, this is a driver of innovation and productivity in other sectors. Second, in many economies, higher education plays a role in generating and diffusing potentially useful research ideas. With effective linkages between higher education, government and business, these may lead directly or indirectly to innovations in socio-economic outputs and processes. There is potential for enhancing productivity and competitive advantage throughout the economy, and for finding new solutions to socio-economic problems.

In addition to these traditionally analysed roles of universities in the development of economies, the HE sector is increasingly recognised more tangibly for its own contribution per se as a potent productive sector, and one with significant spillovers deriving from the consumption activity of students. In particular, it is acknowledged in many countries that the HE sector has potential to generate significant export revenues, both from the attraction of overseas students and from universities establishing a presence in foreign markets. The importance of the sector in this 'market competitive' sense has been present for many years, but has come to the fore recently. For example, Kelly et al. (2006) argue that the growing economic importance of the UK HE sector is reflected in: income of £16.87

billion in 2003/4 (compared with £12.8 billion in 1999/2000); £45 billion of output when multiplier effects are counted; export earnings of £3.6 billion; and the employment of 1.2 per cent of the total UK workforce. Working with a broader industry definition and focusing on exports, Lenton (2007) reports that the total value of education and training exports to the UK economy was almost £28 billion in 2003/4, more than both the financial services (£19 billion) and automotive (£20 billion) sectors.[3] The total value of international students in the HE sector alone was almost £6 billion (Lenton, 2007).

Singapore provides an interesting example of a government that has taken the economic potential of the HE sector extremely seriously. In the late 1990s their Economic Development Board (EDB) launched a project to attract 'world class universities' to Singapore in an explicit and ongoing attempt to establish Singapore as a higher education hub for the Asia Pacific region. As reported on the Institute of International Education website a 2002 government review panel 'set the target of attracting 150,000 foreign students to Singapore by 2015, which it said would not only create 22,000 jobs, but also boost the education sector's GDP contribution from 1.9 percent (US$1.8billion) to 5 percent'.[4] This plan is argued to be on track, with international student numbers rising from below 50000 in 2002 to 72000 in 2005 (www.iienetwork.org/page/116259/).

Returning to the broader contributions of the HE sector, elements of both human capital and research have featured strongly in the development of theoretical and empirical literature seeking to explain economic growth (Lucas, 1988; Romer, 1989; Mankiw et al., 1992; Temple, 1999). Following the pioneering work of Becker (1964), the economic and social returns from higher education are also frequently analysed. The OECD, for example, has integrated such analysis into its influential annual *Education at a Glance* set of indicators (OECD, 2007).[5] More recently, there has been a growing area of research analysing the significance of universities for *local* economic development processes, combining both the human capital and research roles. In particular, a 'third mission' of universities has been identified and progressively explored (Laredo, 2007). Building on traditional functions of educating and conducting research, the importance of the interface between university and local economies has been emphasised in concepts such as 'systems of innovation' (Freeman, 1987; Lundvall, 1992; Nelson, 1993; Cooke et al., 1998), the 'triple helix of industry, government and university' (Etzkowitz and Leydesdorff, 1997) and the 'entrepreneurial university' (Etzkowitz, 1997, 2004). Indeed, combined with a newly invigorated regional focus to analysis of economic development (Storper, 1995; Morgan, 2004), universities are increasingly seen as important actors in transforming territories to 'learning regions' (Morgan, 1997; Feldman, 2001; Lazzeretti and Tavoletti, 2005).

Higher education is also sometimes argued to play an important role in developing capacity for critical thinking among citizens. Barnett (1997), for example, argues that critical thinking is a defining concept of the Western university, developing this into a concept of 'critical being', defined as the ability for people to engage critically with the world and with themselves, developing powers of critical self-reflection and critical action. In this regard there is often a clear distinction made between 'further education' and 'higher education'. While further education tends to provide practical, vocation-specific skills, higher education involves more in-depth 'academic' analysis of a specific subject area, which instead provides more general skills in critical thought and analysis. Another way of thinking of this is that, alongside the provision of certain specific skills and knowledge, higher education creates *a space* for the nurturing and development of creativity in society. In turn this creativity has the potential to spill into other sectors and to impact positively on societies' abilities to innovate in solving a whole range of production and social problems.

In a standard economic argument, the capacity for critical, creative thought might be seen as important in terms of feeding into productive capabilities and innovation processes, enhancing competitive advantage and the ability to solve social problems. In that sense this perspective can be seen as an extension of the various arguments made above regarding the role of universities in economic development processes. However, if we integrate Barnett's (1997) notion of universities stimulating 'critical being' with the strategic choice approach set out in section 2, then we arrive at an alternative perspective on the link between HE and economic development. Specifically, it can be argued that the HE sector plays a potentially significant role in creating 'space' for creative, open 'thought' and 'being', which in turn contributes to the capacity of a society to engage in deliberative processes surrounding its development.

We will go on to suggest that this space for critical thought is being threatened as current trends in the HE sector push towards an over-emphasis on the market competitiveness role. A potential consequence is a deficit of creativity and capacity for critical behaviour in society. We argue that this has implications not only for societal innovation and development in a direct sense, but also for the capacity of societies to engage in deliberative processes around key socio-economic decisions. In particular, therefore, such a deficit has potential to exacerbate the likelihood, scope and depth of strategic failure in a society. We turn now to explore these current trends in the HE sector, using illustrations primarily from UK experience.

4. UNTANGLING CURRENT TRENDS IN THE HE SECTOR

The HE sector in many economies is undergoing fairly dramatic changes associated with what Bok (2003) terms 'commercialisation'. Reflecting on American experience, the previous President of Harvard University and Dean of Harvard Law School argues that this is not an entirely new trend, but rather one that has recently accelerated towards 'unprecedented size and scope' (2003, p. 2). Bok (2003, p. 3) defines commercialisation as 'efforts to sell the work of universities for a profit' and argues that today:

> Opportunities to make money from intellectual work are pursued throughout the university by professors of computer science, biochemistry, corporate finance, and numerous other departments. Entrepreneurship is no longer the exclusive province of athletics departments and development offices; it has taken hold in science faculties, business schools, continuing education divisions, and other academic units across the campus.

The entrepreneurial university is arguably most developed in the United States. The influential Bayh-Dole Act of 1980 granted ownership of intellectual property rights to research institutions, thus creating strong incentives for universities to exploit their research in the marketplace. Over time this interaction with the market has permeated other aspects of universities' operations, from internal management structures and processes to the design and operation of taught academic programmes. Such processes are also at particularly advanced stages in Australia, New Zealand and the UK (Kelsey, 1998; Roberts, 1999) for example, and are spreading rapidly in other advanced economies. This has raised awareness of a variety of problems associated with universities acting increasingly like firms in a market environment (Aronowitz, 2000; Bok, 2003; Grönblom and Willner, Chapter 6 in this volume; Sugden, 2004).

We propose that the evident commercialisation of universities can be usefully untangled into four component processes:

- Marketisation
- Globalisation
- Corporatisation
- Privatisation.

While these are occurring simultaneously and are strongly interrelated, we suggest that their separate identification is helpful in understanding and analysing ongoing trends and their implications for the contribution of universities to socio-economic development processes.

4.1 Marketisation

By 'marketisation' we mean the permeation of market mechanisms in the relationships between higher education and society. We refer specifically to the engagement of universities in markets as 'suppliers' of both 'learning programmes' and 'research'. Hence students and users of research become 'customers' in what is essentially a supplier–client relationship.

In the UK the move towards a market relationship with students can be seen explicitly in the increasing use of market strategies and tools: the introduction of variable 'top-up fees' for undergraduate degree courses; the proliferation of 'self-funded' postgraduate Masters programmes that specifically target different market niches; the use of scholarships and other mechanisms to explicitly price discriminate between students at both undergraduate and postgraduate level. Similar marketisation of the relationship between universities and users of research is evident in the assessment and funding of research activities: internal assessment of research 'value' is increasingly equated with ability to attract funding. This contributes to the development of a clear market for research as academics respond to internal incentives by adopting the role of supplier so as to secure funding from users.

We suggest that the fundamental characteristic of the marketisation process is not the use of prices per se, but rather the change in the nature of the relationship between universities and society: a supplier–client relationship is in essence different to a relationship between two parties with a mutual aim of learning and/or discovery. In this sense we can draw parallels with similar processes in other sectors; with the running of schools and hospitals in the UK, for example (Allen et al., 1999; Allsop and Baggott, 2004).[6] Here, with seeds sewn by the Conservative Thatcher governments of the 1980s, the New Labour Blair governments from 1997 pushed forward the boundaries of marketisation in sectors that were previously considered best sheltered from market relationships. This can be seen in the establishment of 'city academies' and 'foundation hospitals', for example, and the changing nature of the relationship between schools/hospitals and society is evident in the widespread use of rankings and league tables as *pseudo market signals*. Such market signals also abound in the HE sector, alongside other mechanisms such as accreditations. Indeed, the marketisation of the relationship between universities and society is perhaps most clearly reflected in the extreme seriousness with which these signals are treated by universities.

4.2 Globalisation

In the case of higher education the trend of marketisation is occurring alongside an intense *globalisation* of the sector. By globalisation we mean

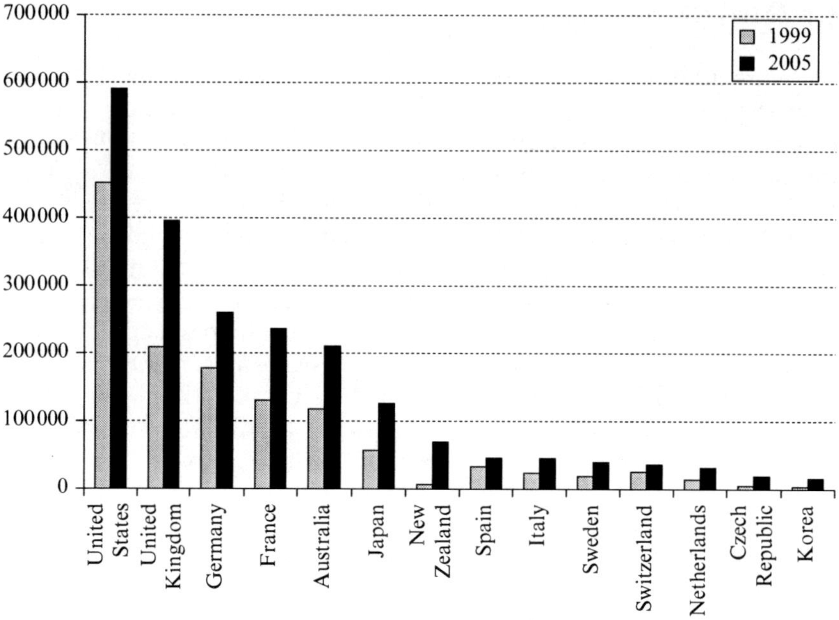

Source: OECD (2007).

Figure 5.1 Growth in international student numbers, 1999–2005

the increasingly global scope of the operations of universities. On the one hand, this can be related to marketisation; globalisation is frequently seen in terms of market relationships (De La Dehesa, 2000). However globalisation has other dimensions, (see, for example: Scholte, 2000; Waters, [1995] 2001). Indeed, in line with a broader approach Sugden and Wilson (2005) suggest that globalisation can be defined as 'a process in which new technologies and new geography imply . . . alterations to economic and associated political, social and cultural relations, behaviour and activity'.

The global scope of HE is not a new phenomenon. For centuries students have travelled to be educated in different countries, but recent years have seen a dramatic acceleration of this trend. Figure 5.1 illustrates the strong growth in numbers of enrolled foreign students in various economies from 1999 to 2005. The trend is clear, with average growth across all of these economies of 66 per cent, or 11 per cent per year. The UK was close to doubling its number of foreign students during this period (88 per cent increase), with almost 400 000 enrolled in 2005. The most striking transformation however has been in New Zealand, where growth has averaged over 150 per cent per year, albeit from a relatively small starting level. The United States meanwhile has exhibited a steadier growth at 5 per cent, but from an already high level – in absolute terms it has also seen an increase in almost 150 000 international students during these six years, which have coincided with a tightening of restrictions for students (and others) entering the country.

Alongside this growth in students travelling to study in other countries, globalisation has seen a new trend as universities themselves have become alert to the opportunities for entering foreign markets, establishing 'branches' in other countries. While this process has been pioneered by US universities, British universities have now also established considerable foreign operations. The Business School of Birmingham University, for example, is one of many UK Business Schools to have established a presence in overseas MBA markets through forms of joint venture.[7] Moreover Nottingham University has led the way in the ambitious approach of establishing own campuses overseas. They have done so in China and Malaysia so as to provide a wide range of undergraduate and postgraduate programmes.[8] In the case of China, the campus is even modelled on that in Nottingham, complete with lake, bell tower and replica of the main administrative building. As we have seen, in the case of Singapore, such 'globalisation' is being encouraged by specific government policy to attract the activities of leading universities from around the world.

4.3 Corporatisation

The two trends of 'marketisation' and 'globalisation' are undoubtedly linked: the increasing mobility of people and firms as globalisation has intensified has opened the eyes of universities to 'profit' opportunities and made them more aware of their market position and potential; and the introduction of market processes domestically has simultaneously contributed to the desire to seek new markets internationally. They are also helping to drive, and being driven by, the evolution of internal corporate structures, practices and principles within universities, a process that we distinguish as 'corporatisation'.

Corporatisation is related to the concept of 'new managerialism' analysed by Lazzerretti and Tavoletti (2006, 24–5) in their discussion of models of governance in higher education. In this model university governance is characterised by 'loose procedural control and tight substantial control with a service belief system'. Universities are said to be 'moving away from the classical Weberian scheme of values and rule observance that is always typical of bureaucracy. They adopt a more entrepreneurial spirit and take advantage of the new huge procedural freedom'. Grönblom and Willner (Chapter 6 in this volume) frame similar arguments in terms of adoption of the 'new public management', defined by Frey and Benz (2005, F377) as 'the view that society should be run as if it were a firm . . . urging non-profit firms and public administrations to adopt pay-for-performance programmes'.

In the case of the UK, the spread of market-oriented principles is occurring across all levels of the internal organisation of universities, alongside the development of more hierarchical management structures. Such corporatisation can be observed specifically in a number of trends: the increasing separation of management from academic activities within universities; the growing use of internal markets organised around distinct 'budget centres'; the evolution of promotion and salary structures to reflect market value, for example through emphasis on attracting funding or developing profitable programmes; the explicit recruitment and retaining of 'star professors' on part-time bases, often seen purely in terms of their value in enhancing market reputation; the rising obsession with internal control mechanisms for issues such as corporate branding and public relations. As argued by Sugden (2004), these types of issues reflect more generally an explicit choice by universities to mimic the structures and behaviour of transnational corporations.

In part corporatisation is a response to increased exposure to domestic and international markets, which generates a perceived need to mimic corporations in order to best take advantage of market opportunities and ward off market threats. However, again there are strong interdependencies. The adoption of corporate principles internally, possibly for efficiency motivations related to the 'new public management' (Frey and Benz, 2005; Pollitt and Bouckaert, 2004), has also helped to facilitate and accelerate the *marketisation* and *globalisation* processes analysed previously. For example, the replacement of (or removal of meaningful power from) academic committee structures enables hierarchical decision-making structures that remove some of the checks and balances in academic decision-making. In turn this makes it easier for market concerns to be expressed at the expense of other potential considerations around the relationship between university and society.

4.4 Privatisation

The final, and perhaps most extreme of the distinct processes that we can identify as part of ongoing commercialisation is that of *privatisation*. By privatisation we mean the explicit transfer of ownership of activities within the HE sector to private, for-profit corporations. While private universities are common and of growing influence in many countries, universities in the UK remain predominantly public institutions operating

Table 5.1 Agreements for private provision of activities with British universities

INTO (UK)	Study Group International (UK)	Kaplan (USA)	IBT Education (Australia)	In-Search (Australia)
University of East Anglia *(2004)*	University of Sussex *(2005)*	Nottingham Trent University *(2005)*	University of Hertfordshire *(2000)*	University of Essex *(2000)*
University of Exeter *(2006)*	University of Stirling *(2007)*	University of Sheffield *(2006)*	Brunel University *(2004)*	
Newcastle University *(2007)*	University of Surrey *(2007)*	University of Glasgow *(2007)*		
	Lancaster University *(2007)*	University of Liverpool *(2007)*		
	University of Huddersfield *(2007)*			
	Liverpool John Moores University *(2007)*			
	Heriot-Watt University *(2007)*			

Source: Websites of the five companies (http://www.into.uk.com; http://www.studygroup.com; http://www.kaplan.com; http://www.libt.uk.com; http://www.insearch.essex.ac.uk) and Newman (2007).

under Royal charter. However, there has been an extremely recent trend for these public universities to transfer significant parts of their learning activities to the private, for-profit sector.

Table 5.1 identifies five private sector education providers of UK, US and Australian origin that have recently established agreements with a range of British universities to provide specific programmes or parts of programmes. In addition to these 17 universities, Birmingham University

(Kaplan), Leeds University (INTO), Leicester University (SGI), London South Bank University (INTO) and Swansea University (IBT) have also recently been reported to have been in talks regarding similar agreements (Newman, 2007). In this early wave of privatisation such agreements have focused on provision of foundation courses prior to entry into the undergraduate and/or postgraduate programmes of the universities concerned, and on provision of English language teaching within the universities. The agreement between Sheffield University and Kaplan International, for example, establishes Sheffield International College to provide two types of qualifications.[9] The Certificate in University Foundation Studies in either 'Business, Law and Social Sciences' or 'Science and Engineering' guarantees direct entry into the first year of undergraduate studies at Sheffield University. The Graduate Diploma in either 'Business, Law and Social Sciences' or 'Science and Engineering' similarly guarantees entry into related postgraduate programmes at Sheffield University.

Again, there are strong links between this privatisation of activities and the processes of marketisation, globalisation and corporatisation. Agreements such as these are attractive to universities precisely because of their embracing of market principles both internally and externally: the private companies bring global marketing expertise and promise to increase admissions, in particular of foreign students, and their control of significant areas of activity enables a streamlining of the internal management hierarchy as groups of staff fall under the new institutional arrangements. The development of foundation programmes themselves also facilitate desired expansion into global markets, easing the burden of evaluating a whole range of different national qualifications and opening a clear route for international students into British universities.

5. AN INTERTEMPORAL TRADE-OFF?

These four interrelated processes have unfolded rapidly in the HE sector in a relatively short space of time, and they continue to advance at pace. Within universities, certainly in the UK, such changes are creating significant tensions, and raising concerns among many academic and administrative staff. Pressures are evident in admissions processes, for example. As admissions tutors increasingly work towards 'targets' for market-oriented programmes, there is pressure to admit students on market rather than academic grounds. Moreover, the privatisation of foundation programmes alongside guaranteed entry for those that complete these programmes brings obvious dangers; there is an effective transfer of admissions control to private companies whose interests are very clearly in ensuring that large numbers of students pass successfully through their programmes and into the university.

Most significantly there is a related change in academic and student perceptions of the learning process as the relationship becomes one of supplier–client. From the student side, while enhanced expectations of standards of teaching and facilities have the potential to drive a positive process of improvements, the implications of the change in relationship are in reality more complex. Alongside positive demands for things such as better e-learning facilities and more effective forms of interaction with academics, 'customer' expectations in the market relationship frequently emerge in terms of belief in a 'right' to a certain degree and to marks above a certain level. Students start to see themselves as purchasing a product, indeed an expensive product. From the academic side there is also a market rationale to honour such a relationship, as failures and low grades have potentially severe market implications in terms of future applications and admissions. Overall this manifests itself (from both sides) in an increasing tendency for students to be taught as 'customers' rather than as people engaging in a learning process, and without great care can easily develop into a spiral of lowering academic expectations.

Similarly in research, pressures to engage in the market for research funding are growing as appointments and promotions increasingly require demonstration of the ability to attract external funding. As these pressures bite, a logical outcome is for academics to stop designing research projects according to their own perspectives on desirable research agendas *and then* seeking the possibility of appropriate funding. Rather, there is a tendency to design and conduct research from the outset according to the explicit (or perceived) objectives of those who fund the research. In the extreme there may even be temptations to manipulate research findings in certain cases, so as not jeopardise future funding opportunities. Again, a change in relationship between academics and the agents that (potentially) use academic research is at the heart of this process. As academics become more concerned with 'supplying' research to 'customers', there is pressure to ensure that the 'product' matches expectations. Such pressures have the potential to fundamentally change the nature of much research; leading to the development a more deterministic, outcome-oriented process that is arguably less free, open-ended and independent.

While these tensions are currently being played out in universities, and boundaries are being explored, we suggest that, crucially, they have potential impacts on the roles that universities are able to play in their society. In particular, there is a strong intuitive argument that the move towards commercial rationale as guiding principles in both learning and research is damaging the capacity of universities to provide the space for independent, critical thought. As students become concerned with achieving the product that they have paid handsomely for, they lose sight of learning as an enriching and reflective process, and that space is invaded. As academics become concerned with meeting their market-determined targets in developing programmes and research, their space to think, reflect and let truly original ideas and agendas emerge is similarly invaded.

There is much discussion within universities around these trends, and detailed research is needed to uncover the nature and magnitude of such impacts on critical thought and creativity. We argue that such research is urgent, as while these tendencies are relatively recent, they are moving fast and have potentially important implications for the long-term economic development of societies. In particular we suggest the possibility of an intertemporal trade-off that has not yet been fully appreciated. The relatively recent desire for higher education to generate *economic rents today*, as the commercialisation of activities is vigorously embraced, may be undermining the capacity of universities to provide space for critical thinking that will impact positively on *economic development processes tomorrow*. There is a danger that lack of awareness of this trade-off is leading us towards a deficit in independent, critical thought and creativity among both graduates and academics.

The potential impacts on long-term economic development from the thin end of this trade-off are significant. They can be seen conventionally in terms of consequences for the productivity and innovative capacity of economies, implying a damaging of market competitiveness. We argue, however, that there is a further, less-appreciated channel through which this trade-off is likely to manifest itself. Following the strategic choice literature, a crowding out of the space for critical, independent thought within universities is likely to undermine the capacity of a society to engage in deliberative, democratic decision-making around economic development processes. In turn the likelihood, scope and depth of strategic failure in society are enhanced.

There are urgent questions regarding the role of policy in addressing the risks highlighted in this chapter. Moreover there are similar questions around what academics working within universities can do to ensure that space for creativity and independent, critical exploration is not crowded-out. In framing this trade-off, the chapter aims to highlight such issues as an important agenda for research, and to provide a reference point for future analysis of different aspects. Indeed, the untangling of 'commercialisation' suggested here is helpful in analysing current reality. It raises questions around the relative significance of each distinct process in terms of overall consequences. To take one of many possibilities, it may be that some of the consequences of 'marketisation' are less severe if 'corporatisation' and/or 'privatisation' processes are tempered, for example. These are the type of detailed issues that require urgent research if the HE sector is to fulfil its potential in making a holistic, balanced and long-term contribution to the economic development of societies.

ACKNOWLEDGEMENT

The author is grateful for helpful discussion and comments on earlier material with Silvia Sacchetti, Roger Sugden and participants at the DARE Workshops in Limerick (2006), Bath (2006) and Reus (2007), at the Biannual Festival of Creativity and Economic Development in Gambettola (2007), and at the Association of Social Economics World Forum in Amsterdam (2007).

NOTES

1. See, for example, Baran and Sweezy (1966), Hymer (1970), Cowling (1982), Encaoua et al. (1986), Tirole (1988), Geroski (1995) and Cowling and Sugden (1998).
2. Consider also Fromm's (1941) view that freedom can be accompanied by uncertainties and fears that lead people to seek an escape. In 'free market' economies, where the mechanism of 'exit' predominates, Fromm contends that people have 'freedom from constraints' without having 'freedom to do'. Their reaction can be to turn their backs on freedom, accepting conformity and the decisions of others. Such a phenomenon might help to explain, for example, the downward trend in voter turnout in Western economies over recent decades.
3. See 'Education worth more to British exports than banking', *The Guardian*, 18 September 2007.
4. www.iienetwork.org/page/116259.
5. See also Blundell et al. (2000) for an example of an econometric analysis using UK data; Psacharopoulos (1994) or Psacharopoulos and Patrinos (2002) for updates of key findings regarding returns to investment in education.
6. See also Ainley (2004) on the new 'market state' with regards the education sector in the UK.
7. See http://www.business.bham.ac.uk/postgraduate/mba/int_business/global.shtml.
8. See www.nottingham.edu.cn; www.nottingham.edu.my.
9. See www.sheffield.ac.uk/sic/programmes.

REFERENCES

Ainley, Patrick (2004), 'The new "market-state" and education', *Journal of Education Policy*, **19**(4), 497–513.

Allen, Martin, Caroline Benn, Clyde Chitty, Mike Cole, Richard Hatcher, Nic Hirtt and Glen Rikowski (1999), *Business Business Business: New Labour's Education Policy*, London: Tufnell Press.

Allsop, Judith and Rob Baggott (2004), 'The NHS in England: from modernisation to marketisation?', in N. Ellison, L. Bauld and Martin Powell (eds), *Social Policy Review 16: Analysis and Debate in Social Policy, 2004*, Bristol: Policy Press.

Aronowitz, Stanley (2000), *The Knowledge Factory: Dismantling the Corporate University and Creating True Higher Learning*, Boston, MA: Beacon Press.

Bailey, David, Lisa De Propris, Roger Sugden and James R. Wilson (2006), 'Public policy for economic competitiveness: an analytical framework and a research agenda', *International Review of Applied Economics*, **20**(5), 555–72.

Baran, Paul A. and Paul M. Sweezy (1966), *Monopoly Capital: An Essay on the American Social and Economic Order*, New York: Monthly Review Press.

Barnett, R. (1997), *Higher Education: A Critical Business*, Bristol: Open University Press.

Becker, Gary S. (1964), *Human Capital: A Theoretical and Empirical Analysis, with Special Reference to Education*, New York: National Bureau of Economic Research.

Blundell, Richard, Lorraine Dearden, Alissa Goodman and Howard Reed (2000), 'The returns to higher education in Britain: evidence from a British cohort', *Economic Journal*, **110**(461), F82–F99.

Bohman, James (1998), 'Survey article: the coming age of deliberative democracy', *Journal of Political Philosophy*, **6**(4), 400–25.

Bok, Derek (2003), *Universities in the Market Place: The Commercialisation of Higher Education*, Princeton, NJ: Princeton University Press.

Branston, J. Robert and James R. Wilson (2006), 'Transmitting democracy: a strategic failure analysis of broadcasting and the BBC', University of Bath School of Management working paper no. 2006.11.

Branston, J. Robert, Keith Cowling and Roger Sugden (2006a), 'Corporate governance and the public interest', *International Review of Applied Economics*, **20**(2), 189–212.

Branston, J. Robert, Lauretta Rubini, Roger Sugden and James R. Wilson (2006b), 'The healthy development of economies: a strategic framework for competitiveness in the health industry', *Review of Social Economy*, **LXIV**(3), 301–29.

Branston, J. Robert, Roger Sugden, Pedro Valdez and James R. Wilson (2006c), 'Generating participation and democracy: an illustration from electricity reform in Mexico', *International Review of Applied Economics*, **20**(1), 47–68.

Coase, Ronald H. (1937), 'The nature of the firm', *Economica*, **4**(16), 386–405.

Coe, Neil M. and Henry W. Yeung (2001), 'Geographical perspectives on mapping globalisation', *Journal of Economic Geography*, **1**, 367–80.

Cooke, P., M. Gomez Uranga and G. Etxebarria (1998). 'Regional systems of innovation: institutions and organisational dimensions', *Research Policy*, **26**(4–5), 475–91.

Cowling, Keith (1982), *Monopoly Capitalism*, London: Macmillan.

Cowling Keith and Roger Sugden (1987), *Transnational Monopoly Capitalism*, Brighton: Wheatsheaf Books.

Cowling, Keith and Roger Sugden (1998), 'The essence of the modern corporation: markets, strategic decision-making and the theory of the firm', *Manchester School*, **66**(1), 59–86.

Cowling, Keith and Roger Sugden (1999), 'The wealth of localities, regions and nations: developing multinational economies', *New Political Economy*, **4**(3), 361–78.

Cowling, Keith, Silvia Sacchetti, Roger Sugden and James R. Wilson (2009), 'The United Nations and globalisation: a reconnaissance of the issues', in J. Davis (ed.), *Global Social Economy: Development, Work and Policy*, Oxford: Routledge.

Curran, James (1991), 'Rethinking the media as a public sphere', in: P. Dahlgren and C. Sparks (eds), *Communication and Citizenship: Journalism and the Public Sphere in the New Media Age*, London: Routledge.

De La Dehesa, Guillermo (2000), *Comprender la Globalización*, Madrid: Alianza Editorial.

Dewey, John (1927), *The Public and its Problems*, reproduced in J.A. Boydston (ed.) (1988), *John Dewey. The Later Works Volume 2: 1925–1927*, Carbondale, IL and Edwardsville, IL: Southern Illinois University Press.

Elster, Jon (ed.) (1998), *Deliberative Democracy*, Cambridge: Cambridge University Press.

Encaoua, David, Paul Geroski and Alexis Jacquemin (1986), 'Strategic competition and the persistence of dominant firms: a survey', in: J.E. Stiglitz and G.F. Mathewson (eds), *New Developments in the Analysis of Market Structure*, London: Macmillan.

Etzkowitz, H. (1997), 'The entrepreneurial university and the emergence of democratic corporatism', in H. Etzkowitz and L. Leydesdorff (eds), *Universities and the Global Knowledge Economy: A Triple Helix of University–Industry–Government Relations*, London: Cassell.

Etzkowitz, H. (2004), 'The evolution of the entrepreneurial university', *International Journal of Technology and Globalisation*, **1**(1), 64–77.

Etzkowitz, H. and L. Leydesdorff (eds) (1997), *Universities and the Global Knowledge Economy: A Triple Helix of University–Industry–Government Relations*, London: Cassell.

Feldman, J.M. (2001), 'Towards the post-university: centres of higher learning and creative spaces as economic development and social change agents', *Economic and Industrial Democracy*, **22**(1), 99–142.

Freeman, C. (1987), *Technology Policy and Economic Performance. Lessons from Japan*, London: Pinter.

Frey, Bruno, S. and Matthias Benz (2005). 'Can private learn from public governance', *Economic Journal,* **115**(507), F377–F396.

Fromm, Eric (1941), *Escape from Freedom*, 1999 edn, New York: Owl Books.

Geroski, Paul (1995), 'What do we know about entry?', *International Journal of Industrial Organisation*, **13**, 421–40.

Habermas, Jürgen (1964), 'The public sphere', reprinted in P. Golding and G. Murdock (eds) (1997), *The Political Economy of the Media, Volume II*, Cheltenham, UK and Lyme, USA: Edward Elgar.

Hirschman, Albert O. (1970), *Exit, Voice and Loyalty: Responses to Decline in Firms, Organizations and States*, Cambridge, MA: Harvard University Press.

Hymer, Stephen (1970), 'The efficiency (contradictions) of multinational corporations', papers and proceedings, *American Economic Review*, **60**(2), 441–8.

Kelly, Ursula, Donald McLellan and Iain McNicoll (2006), *The Economic Impact of UK Higher Education Institutions: A Report for Universities UK*, Glasgow: University of Strathclyde.

Kelsey, J. (1998), 'Privatizing the universities', *Journal of Law and Society*, **25**(1), 51–70.

Laredo, P. (2007), 'Revisiting the third mission of universities: toward a renewed catergorization of university activities', *Higher Education Policy*, **20**, 441–56.

Lazzeretti, L. and E. Tavoletti (2005), 'Higher education excellence and local economic development: the case of the entrepreneurial university of Twente', *European Planning Studies*, **13**(3), 475–93.

Lazzeretti, Luciana and Ernesto Tavoletti (2006). 'Governance shifts in higher education: a cross-national comparison', *European Educational Research Journal*, **5**(1), 18–37.

Lenton, Pamela (2007), *Global Value. The Value of UK Education and Training Exports: An Update*, London: British Council.

Long, Norton E. (1990), 'Conceptual notes on the public interest for public administration and policy analysts', *Administration Society*, **22**(2), 170–81.

Lucas, Robert E. Jr. (1988), 'On the mechanics of economic development', *Journal of Monetary Economics*, **22**, 3–42.

Lundvall, B.-A. (1992), *National Systems of Innovation. Towards a Theory of Innovation and Interactive Learning*, London: Pinter.

Mankiw, Gregory N., David Romer and David N. Weil (1992), 'A contribution to the empirics of economic growth', *Quarterly Journal of Economics*, **107**(2), 407–37.

Morgan, K. (1997), 'The learning region: institutions, innovation and regional renewal', *Regional Studies*, **35**(5), 491–503.

Morgan, K. (2004), 'The exaggerated death of geography: learning, proximity and territorial innovation systems', *Journal of Economic Geography*, **4**, 3–21.

Nelson, R.R. (ed.) (1993), *National Systems of Innovation*, Oxford: Oxford University Press.

Newman, M. (2007), 'Private firms tighten grip', *Times Higher Education Supplement*, 20th April.

Organisation for Economic Co-operation and Development (OECD) (2007). *Education at a Glance 2007: OECD Indicators*, Paris: OECD.

Pollitt, Christopher and Geert Bouckaert (2004), *Pubic Management Reform: A Comparative Analysis*, 2nd edn, Oxford: Oxford University Press.

Psacharopoulos, George (1994). 'Returns to investment in education: a global update', *World Development*, **22**(9), 1325–43

Psacharopoulos, George and Harry A. Patrinos (2002). 'Returns to investment in education: a further update', World Bank policy research working paper no. 2881, Washington, DC.

Roberts, P. (1999), 'The future of the university: reflections from New Zealand', *International Review of Education*, **45**(1), 65–85.

Romer, Paul M. (1989), 'Human capital and growth: theory and evidence', National Bureau of Economic Research working paper no. 3173, Cambridge, MA: NBER.

Sacchetti, Silvia and Roger Sugden (2003), 'The governance of networks and economic power: the nature and impact of subcontracting networks', *Journal of Economic Surveys*, **17**(5), 669–91.

Sacchetti, Silvia and Roger Sugden (2007), 'Creativity in economic development: space in an inferno', mimeo, Institute for Economic Development Policy, University of Birmingham.

Scholte, Jan Aart (2000), *Globalization: A Critical Introduction*, Basingstoke: Palgrave.

Sugden, Roger (2004), 'A small firm approach to the internationalisation of universities: a multinational perspective', *Higher Education Quarterly*, **58**(2–3), 114–35.

Sugden, Roger and James R. Wilson (2002), 'Development in the shadow of the consensus: a strategic decision-making approach', *Contributions to Political Economy*, **21**, 111–34.

Sugden, Roger and James R. Wilson (2005), 'Economic globalisation: dialectics, conceptualisation and choice', *Contributions to Political Economy*, **24**, 1–20.

Storper, M. (1997), *The Regional World: Territorial Development in a Global Economy*, London: Guildford Press.

Temple, Jonathan (1999), 'The new growth evidence', *Journal of Economic Literature*, **37**(1), 112–56.

Tirole, Jean-Jacques (1988), *The Theory of Industrial Organization*, Cambridge, MA: MIT Press.

Warren, Mark (2005), 'Deliberative democracy', in A. Carter and G. Stokes (eds), *Democratic Theory Today*, Cambridge: Polity Press.

Waters, Malcolm (2001), *Globalization*, 2nd edn, London: Routledge.

Wilson, James R. (2004), 'Strategic decision-making in development theory and practice: a learning approach to democratic development', PhD thesis, University of Birmingham, (http://etheses.bham.ac.uk/115/).

Zeitlin, Maurice (1974), 'Corporate ownership and control: the large corporation and the capitalist class', *American Journal of Sociology*, **79**(5), 1073–119.

PART III

Emerging institutional settings, critical thinking and knowledge

6. Destroying creativity? Universities and the new public management

Sonja Grönblom and Johan Willner

1. INTRODUCTION

In this contribution we analyse the university as an economic organisation, mainly because university reforms are advocated with economic arguments and not because of any alleged superiority of economics in understanding the nature of research and higher education. Budget cuts, managerialism, performance management with a focus on output, and often economic sticks and carrots at all levels seem to be the order of the day almost everywhere. We shall therefore ask whether such reforms really make economic sense, and make a case for an autonomous university characterised by collegiality and academic freedom.

This chapter is organised as follows. The next section briefly describes the so-called new public management and its consequences for universities. Section 3 warns against false analogies and describes the traditional university. Section 4 focuses on incentives, because of the underlying belief that university employees would not work properly without sticks and carrots. We make the point that the role of incentives cannot be to make academics work harder, because most of them are highly motivated. The following section (5) puts the incentives into context as a mechanism for motivation-crowding out. This is a prerequisite for a strong leadership that redirects the university so as to focus on product and process innovations for the business community. Section 6 includes some complementary comments and section 7, concluding remarks.

2. THE NEW PUBLIC MANAGEMENT

Those who advocate university reforms often refer to inefficiency (see, for example, Jacobs and van Ploeg, 2006). Universities are not unique in this respect. The so-called 'new public management' is based on the view that the public sector tends to be inefficient and is in need of a shake-up. Similar reforms as in the universities are therefore affecting most public-sector activities, not only public administration but public production such as in health care, research and education and other activities where complete privatisation is not an option.

The new public management is not the outcome of a coordinated conspiracy or a unified blueprint. Some authors trace its origins back to the Californian tax revolts in the 1970s (Gruening, 2001). Thatcherism in the UK and the economic reforms in New Zealand and Australia are other well-known examples of policies that later came to be described in terms of the new public management. The OECD is a strong advocate (Lawton, 2005), and policies in the spirit of the new public management are now also applied in the Netherlands, Germany and the Nordic countries (Gruening, 2001, Henriksson, 2004, Schimank, 2005, Huisman et al., 2006).

Hood (1995) describes the new public management as a reaction against the so-called 'progressive public administration', which emphasised the difference between the public and the private sector and the necessity to avoid corruption. It was associated with rules on inputs (process accountability) rather than payment by result. The new public management became a generic label for the common denominators in shifts that instead, emphasised the private sector as a role model (Hood, 1995; Newberry and Pallot, 2004).

As for its theoretical sources of inspiration, the new public management is usually associated with a particular kind of economic thinking (Frey and Benz, 2005). Its proponents have been inspired by the mainstream principal–agent model, which focuses on work incentives under asymmetric information, but also by the Chicago School and its emphasis on smoothly working markets and the Public Choice School with its assumptions about selfish public-sector decision-makers (Hood, 1995; Henriksson, 2004).

The details of public-sector reform differ from country to country, without too much sign of a convergence (Lawton, 2005).[1] It is harder to understand the ideological background in Finland and other Nordic countries, although the major political parties tend to see a reduction in public expenditure as an end in itself there also. But the argument is often pragmatic: it is argued that a small open economy would lose its international competitiveness unless it follows the same low-tax policy as other countries. Nevertheless, we need a label for the reforms that affect both universities and other public-sector institutions, and the 'new public management' seems suitable.

Most observers associate the new public management with, among other things, budget cuts, vouchers and user charges, accountability for performance and auditing, freedom to manage (and hence top-down management and lost professional autonomy so that weak performers can be punished), and personnel management with an emphasis on economic incentives (Hood, 1995; Gruening, 2001; Schimank, 2005).

In universities, this usually means a stronger central administration and less scope for democratic decision-making, and sticks and carrots for institutions, faculties/schools, departments and individuals. The university budgets are therefore often linked to earlier results, especially in the form of diplomas or exams (i.e. output funding). Moreover, incentive wages are replacing the fixed salaries of the old system, and the role of tenured staff is diminishing. For example, nearly two-thirds of US academic staff are non-tenured (Scheuerman and Kriger, 2004), without the academic freedom for which a permanent position is a precondition.

Some authors argue that the new public management has not (yet) been able to reshape the public sector completely (Lawton, 2005) and that the academics are still in a powerful position (Huisman, de Boer and Goedgebuure, 2006). But this makes it all the more relevant to make a critical assessment of the new public management as a doctrine for university reforms that may be difficult to reverse.

3. WHAT IS A UNIVERSITY?

The new public management represents just one of several analogies that outsiders have used to understand universities. Universities have earlier been compared to clerical organisations, schools (in primary or secondary education), research institutes, or departments of the public administration. The late 1960s and the 1970s tended to see the university as part of a political system, and a movement in favour of one man, one vote became strong, particularly in Finland. But now the business enterprise seems to be the standard of comparison.

While a university may have a superficial resemblance to many other types of institutions, it has distinctive features of its own. It is not a school, despite the fact that most of its staff are involved in teaching, because higher education requires the teacher to be a researcher. A university is not like governmental or other research institutes that produce commissioned reports, despite the fact there is usually also some commissioned research in a university. The main part of the university research has traditionally been initiated by its own academic staff and free from intervention from decision-makers in the public sector or the business community (academic freedom). Universities are not government departments although many universities, and in Europe most of them, arc owned by the state.

Finally, a university is not like a firm, despite the new public management's insistence on business-like practices. Students contribute to the funding of universities to varying degrees (and there are still no tuition fees in Scandinavia). But price-to-cost ratios tend to be on average 0.067 also among the top public universities in the US, and they are no higher than (on average) 0.89 among the bottom private institutions (Winston, 1999). More fundamentally, most of the basic and applied research that is initiated by the academic research cannot be sold to paying customers. The task of university researchers is partly to initiate research that no outsider would commission, and often, in particular, such research that would be an anathema to those who can afford to pay for commissioned research.

A university has, of course, a superficial resemblance to these and many other types of organisations, and we want in particular to emphasise so-called knowledge-based firms, like consulting firms, architectural agencies, medical centres or newspapers, which are led by experts in their fields rather than by professional administrators. An editor-in-chief should for example be a qualified journalist, able to write editorials. Bosses without expertise in the activity of the organisation would under such circumstances not be respected. The activity must instead be based on creative cooperation (Sveiby and Risling, 1987).

Schimank (2005) describes the traditional university as a combination of political regulation and professional self-control. While the university as a whole is not necessarily autonomous in relation to the political decision-makers in, for example, Scandinavia or Germany, the individual professors are highly independent. Professors have job security and are sometimes derisively described as small businessmen or monopolists that cannot go bankrupt. There is egalitarianism among the professors, and a tradition of collegiality (*Kollegialität*) which means implicit non-aggression pacts and hence a strong emphasis on consensus. The strong leadership prescribed by the new public management is therefore seen as a way to overcome the resistance against reforms that such a structure implies.[2]

However, if a university is a multidisciplinary expert organisation for independent research and teaching, how can a professional leader have the necessary detailed knowledge and skills of all its disciplines? A leader of a single-purpose knowledge-based organisation, such as an editor-in-chief, can have that kind of undisputed authority, but not a vice-chancellor or even a dean or department head.[3] The participatory democracy that characterises the traditional university therefore makes sense and explains why it can even be described as a producer cooperative where nobody is in charge. (Many knowledge-based private-sector companies can also be described as producer-cooperatives; see Sveiby and Risling, 1987).[4]

There is another strong reason for the professional autonomy that characterises the traditional university. How can research be truly innovative if the researchers lose the right to set their research agenda? Basic research can also subsequently turn out to be of fundamental importance for economically useful R&D. Outside intervention in favour of only immediately useful research can stop this process. Moreover, research can be controversial among decision-makers, as in the case of evolutionism in the USA. Economics and other social sciences might also be in the risk zone. Our own university department takes pride in research on income distribution, social mobility and poverty, public and private production (including critical assessments of the new public management). We are not sure that outside powers would always like what we are doing.[5]

The university must therefore be characterised by academic freedom. The tenure of the senior staff is a precondition, because without tenure a researcher would be dependent on superiors. The tenure system in universities can therefore be described as rational for the same reason as the judges have to be independent, and seems to have a strong survival value in the otherwise highly competitive US system (McPherson and Shapiro, 1999). Another change is the diminished role of the tenured staff. For example, nearly two-thirds of US academic staff are non-tenured (Scheuerman and Kriger, 2004). However, we would also describe academic freedom, and hence tenure, as a cornerstone in a civilised society. It sets an example that other organisations in both the public and private sector might follow.

Academic freedom also means that a researcher cannot have superiors in the usual sense, whatever deans and vice-chancellors choose to call themselves. Decisions on teaching are usually made by department or faculty boards, and no authority can decide on behalf of the researchers on whether the earth moves around the sun or vice versa. Managerialism would change the department heads, deans and vice-chancellors into bosses as defined by organisation theory, where a boss supervises and monitors a subordinate's performances (Hess, 1983). A boss determines the salary, contributes to decisions on promotion or dismissal and, above all, decides on daily assignments and all issues where the employment contract and/or the job description are inconclusive.

4. STICKS AND CARROTS

4.1 The Economic Man

Economic incentives are a fundamental part of the new public management. In mainstream economic theory, these are analysed within the principal–agent theory, where agents (shop-floor employees, but more often CEOs and fund managers) are assumed to be opportunistic. There is asymmetric information, so the principal is unable to distinguish between the effects of high or low efforts and random shocks, and this leaves room for selfish maximisation of utility functions that make the agents essentially greedy and lazy: utility is increasing in income but decreasing in effort (or increasing in managerial slack). In politics, such opportunism means that actions are taken for the purpose of being re-elected and not because of idealism. Civil servants are believed to be motivated only by salaries, perquisites and big budgets. When applied on university management, the approach means that researchers and teachers would take the inability of the university bureaucrats to monitor them as an opportunity for laziness.

This kind of lazy and self-interested employee is just an aspect of the conventional economic man (*homo economicus*) in the workplace. Economic models must be based on simplistic behavioural assumptions, and the economic man has been useful for understanding consumer demand or oligopolistic markets, but may be less useful in other areas. For example, a person may be a dedicated employee and not at all lazy and opportunistic, but may nevertheless behave approximately as the economic man in the supermarket. However, arguments based on economic self-interest are often applied *in absurdum*, for example when analysing the choice of spouse or for understanding political and administrative behaviour (for a summary, see for example Lepage, 1980).

In the spirit of the new public management, the university is believed to need economic incentives (sticks and carrots) in order to motivate its researchers, in the absence of competition as a disciplinary force (see, however, section 5). Individual negotiations are replacing the fixed salaries of the old system. Incentives are also introduced for institutions, faculties/schools and departments in some countries. This can mean that university budgets are linked to earlier results, especially in the form of diplomas or exams (i.e. output funding).[6] Since the budget is given in the short run, the system then becomes a zero sum game where universities, departments and individuals are forced to compete in a way that may jeopardise the cooperation that is necessary for successful research.

4.2 Multidimensional Efforts and Short-term Decisions

No employee would provide any effort under asymmetric information without economic incentives in the simplest of the principal–agent models. However, many of them were constructed in order to demonstrate that economically rational behaviour is consistent with market failures (sticky wages or prices or markets with positive or negative excess demand). A price cannot at the same time equate supply and demand and provide incentives or signals or work as a self-selection device.[7] These simple models were meant to bridge a gap between microeconomics and Keynesian macroeconomics, and should not be read normatively, as blueprints for how public-sector organisations should be designed.

As will be argued below, the basic assumption on opportunism can be challenged. However, the simple models can also be amended without questioning their basic behavioural assumptions. For example, members of a producer cooperative (for example a traditional university) tend to work harder if they expect that other employees also do their best, as when they take into consideration each others' reactions instead of behaving like Cournot-competitors. Such a setting has been used to explain why kibbutzim in Israel, without performance-related pay, were in general more efficient than kolkhozes in the Soviet Union, where incomes were performance-related, contrary to what is often believed (Guttman and Schnytzer, 1989).

A simple principal–agent model usually deals with a business enterprise with a well-defined product, so that it is easy to understand what the effort means. A typical academic is on the other hand engaged in several tasks, usually at the same time: solving new research problems, writing reports on the basis of new research, writing overviews, lecturing, keeping lectures up to date, in addition to administrative tasks with varying degrees of necessity. The efforts are in other words multidimensional, whereas simple models describe efforts through a scalar.

Also, some aspects of our work may be essential for the quality of our institutions, but have a limited impact on the measurable output in the short run. In particular, if the amount of diplomas is decisive for the funding of a university department, as in Finland, the measurable output does not increase by doing research and keeping lectures up to date, so our efforts have different effects in the long and short run.

If the principal–agent analysis is amended so as to include multidimensional efforts that have an impact with differing time horizons, it turns out that rewards and penalties that depend only on those efforts that can be observed in the short run can lead to distortion (Holmström and Milgrom, 1991). A business enterprise may for example require research and development which cause just costs and no benefits in the short run. This might explain why many large American companies can actually be described as incentive-free zones: the top salaries are not related to measurable performance (Jensen and Murphy, 1990; Holmström and Milgrom, 1991). In a university, output funding in terms of diplomas can, for example, force the academic staff to neglect such necessary activities that do not affect the number of diplomas in the short run. If output funding also takes refereed articles into account, there is a risk that the researchers produce mostly spin-offs along well-trodden paths rather than becoming engaged in risky projects.

Multidimensionality and the presence of certain kinds of efforts that have visible consequences only in the very long run are essential features in public-sector institutions and particular in universities. It is therefore unfortunate that the reform agenda has become inspired by highly simplistic versions of economic thinking that were in addition designed for completely different purposes.

4.3 Intrinsic Motivation

The scope for higher efficiency through an incentive structure that yields higher individual efforts would of course be limited if the academic staff are dedicated rather than lazy. Any positive effects might in addition be overshadowed by the fact that performance-based pay also requires assessment and hence costs in terms of time in the case of each employee and money in the case of the employer. Forms have to be filled in, and there has to be negotiations if wages are set individually. There must be bureaucrats that read through the forms and make decisions on the wage of each employee.

To assert that all politicians and administrators are opportunists (or the opposite) is an ideological statement which is contradicted by the experiences of daily work within a university. In the literature, public-sector employees are often described as prepared to provide efforts because they value for example a social service (a public service motivation).[8] Public (or non-profit) provision is therefore not controversial if employees are mission-oriented or characterised by a public-sector ethos (Francois, 2000; Besley and Ghatak, 2005). Academics are hardly an exception. A new consensus is emerging on the fact that economic agents cannot always be assumed to behave like the economic man (Fehr and Falk, 2002).

Intrinsic motivation means that an employee derives utility and not only costs from efforts as such or from the organisation's performance (Frey, 1997). A significant part of the literature on intrinsic motivation has dealt with commercial enterprises rather than universities and other non-profit organisations. A telephone survey of almost 1700 full- or part-time employees in various companies in the USA suggests for example that most respondents (82.7 per cent) were willing to work according to what was agreed with the employer, even without monitoring (Minkler, 2004). The respondents may have exaggerated their motivation, but the result is on the other hand consistent with experimental studies suggesting that many employees are characterised by reciprocity: an employer that is perceived as honest makes the employees honest (Fehr and Fischbacher, 2002; Fehr and Falk, 2002). But intrinsic motivation is likely to be even more significant for academic work where employees can be driven by curiosity. If so, shirking is not such a problem as suggested by a careless reading of the principal–agent literature. This observation seems to apply for universities in particular, because of the prominent role of intrinsic motivation.

An example with similarities with university research relates to the Open Source Software (or OSS). The OSS is developed by relatively young, motivated and highly educated persons and is made available free of charge like a public good. The economic motive cannot be prominent, since their work is not likely to yield pecuniary rewards. The primary driving force is the perception that better software is needed. This would suggest that the producers are driven by intrinsic motivation, or by the play value or *homo ludens* payoff of their activity, and a willingness to make a donation to society (Bitzer et al., 2004).

However, an economy cannot rely on only such voluntary work as in the OSS community, so there have to be non-profit organisations that pay an appropriate salary in areas where commercial provision is not possible or desirable. We have analysed a university where employees are driven by intrinsic motivation elsewhere (Willner and Grönblom, 2009). The model in question is based on a multiple self as described in Elster (1986): one side of the agent's personality is treated as dedicated to work whereas another maximises expected utility as described in traditional principal–agent theory. The outcome of the bargaining game between the two sides of the personality is affected by, for example, the wage schedule. We also assume that performance-related pay requires the creative staff to do some amount of bureaucratic work. There is then only disutility but no intrinsic motivation. There is a corresponding cost for the employer as well, who has to appoint a number of bureaucrats that read all forms that the academic staff are required to fill in.

The analysis suggests that efforts and hence output per employee can indeed be higher when the wage is fixed than when there is performance-related pay under reasonable circumstances. This happens if the intrinsic motivation is strong, and if there are sufficiently high costs for employer and employee caused by the monitoring and bureaucracy that is necessary for implementing such a wage structure. (The subjective disutility is what matters for the employee, and this can be quite high for creative individuals!) It also turns out that a cut in the wage budget reduces the effort and output per capita, which are also higher when the budget is fixed than in the presence of output funding.

The employer is in conventional models constrained by the fact that the employee would quit if the employment contract is unsatisfactory (the participation constraint). As this constraint is in general binding, the presence of some intrinsic motivation would reduce the wage, because less is needed to keep the agent who enjoys working satisfied.[9] However, the constraint is not always binding in the model with split personalities, so it follows that motivated agents can get higher wages under some circumstances.

5. LEADERSHIP AND MOTIVATION CROWDING-OUT

It has been suggested that economic incentives may crowd out the intrinsic motivation or decrease the employee's ability to perform (Frey, 1997; Fehr and Falk, 2002; Minkler, 2004). The employee might be burned out, perceiving that the job provides no satisfaction in itself but only the means to survive. Such crowding-out of the intrinsic motivation reminds of the concept of alienation in Marxian theory.

Bénabou and Tirole (2003) explain motivation crowding-out by assuming that employees are assumed to have incomplete information about their capacity and the character of their assignment. The agent's behaviour is affected partly by economic payoff, but also by how the employer intervenes. Such factors affect the employee's self-image and the extent to which her assignments are perceived as qualified. This may be of great importance for her self-confidence and hence for her intrinsic motivation (Bénabou and Tirole, 2002). The employer has often a vested interest in assigning difficult tasks to an employee and may even manipulate information in order to improve her self-confidence.[10] Payments that are unrelated to performance signal confidence, hence strengthening intrinsic motivation, whereas rewards and punishments can make a task seem less attractive. They might encourage the employee in the short run, but in the long run they reduce the willingness to exert efforts.

James (2005) represents another but related approach to understand motivation crowding-out. The model works as if employees choose to be either more or less dedicated to their work. Non-monetary rewards can then increase efforts by signalling appreciation or recognition, whereas sticks and carrots can reduce the employee's engagement, in particular if the scheme is perceived as controlling and if the innate intrinsic motivation is weak.[11] This approach combines a Benthamite notion of utility as a sum (which can be maximised with different degrees of dedication to the work) with insights from cognitive evaluation theory.[12]

Our companion paper on multiple selves and intrinsic motivation in a university (Willner and Grönblom, 2009) implies a slightly different mechanism of motivation crowding-out. If the participation constraint is binding, performance-related pay makes the employee behave as if there was no intrinsic motivation: it becomes crowded out as when a candle is blown out. This effect is not caused by the economic incentives per se, but by the participation constraint that keeps the employee at the same utility level irrespective of effort. But it also turns out that this constraint is unlikely to be binding in the presence of reasonably strong intrinsic motivation, for the simple reason that enjoying a work assignment can compensate for a low wage. It would therefore appear at first hand that intrinsic motivation can survive the new public management.

However, it also turns out that there can be motivation crowding-out when the participation constraint is not binding. The employer can choose the optimal bonus but set the constant part of the wage at such a low level that the employee is near the limit of quitting. This has the same impact on intrinsic motivation. To set the wage in this way would not be strictly rational in the light of the technical analysis, but we may ask whether there are aspects that are not captured by our formal analysis that might explain such behaviour.

To understand why seemingly irrational motivation crowding-out might occur, note that intrinsic motivation is not always beneficial for the employer (Frey, 1997). Intrinsic motivation can for example make employees more opinionated and hence less cooperative than those who are driven by economic incentives and hence can be bought. The postal system and the railways are not usually run by philatelist or steam-engine enthusiasts. But a university should presumably value precisely this kind of dedication.

However, the new public management also favours managerialism. The traditional university is seen as hostile to such reforms that would adapt the university to objectives set by non-academics, such as businessmen and politicians who value product and process innovations but not social criticism. To reorganise universities towards economically useful disciplines would require measures to downgrade or eliminate those disciplines that are perceived as irrelevant or costly and possibly weak. This is inconsistent with the traditional university's traditions of consensus decisions. The reforms cannot therefore be implemented without a stronger leadership, but the informal pressure from senates and faculty councils with a reduced formal power can even then be subject to considerable informal pressure (Schimank, 2005).

A seemingly irrational management approach, such as making employees feel that the work assignments are less attractive, can be part of a power struggle (ego-bashing), where the employer deliberately undermines the self-confidence of the employee (Bénabou and Tirole, 2003).[13] The employer can also conceal crucial information or underrate the employee's work assignments. In the first case, the damage can be repaired by providing information later, but the damage is irreversible in the latter case (Bénabou and Tirole, 2003).

While dedicated researchers can be compared to artists and other creative professionals, a university bureaucrat or civil servant at the Ministry of Education might prefer precisely those reforms that crowd out the intrinsic motivation and humiliate the academic staff in order to reduce the assertiveness that comes with intrinsic motivation. If academics become less dedicated, they would presumably care less about the future of their departments. A non-dedicated academic whose discipline becomes downgraded can easily be compensated through prestigious administrative assignments or a generous pension scheme. The reforms can then be implemented despite what is left of the consensus culture.

6. OTHER ASPECTS OF THE NPM: COMPETITION AND BUREAUCRACY

If every member of the academic staff were to be unorganised and work on a freelance basis, they would sell their labour hours on a spot market with competition at least on the supply side. But both public-sector organisations and private firms can be characterised as islands of economic planning in the market sea. Their employees are subject to organisational rules rather than market solutions. Since the new public administration is based on the belief that they are greedy and lazy and lack the discipline associated with competition on a market, they become subject to rewards and punishments.

Competition is nevertheless of crucial importance in a present-day university, although the working hours of its tenured staff are not sold on a spot market, and it can even be argued that this constitutes excessive high-powered incentives. If a university is to resemble a company, there will be at least as much competition between individuals about positions as there will be between universities and their units.[14]

However, the budget cuts that have affected the universities in many countries and sometimes the oversized targets for doctoral degrees creates a competition for permanent posts that may be at least as intense as in the business community.[15] Short-term contracts are in Finland at least more widespread in the public sector and in particular in universities than in the private sector. Members of the senior staff must spend a significant part of their time to compete over funds in order to employ young researchers, who are in turn forced to compete over small grants or short assignments. Such uncertainty may affect the research output and sometimes even the physical and mental health of the staff. Competition can, in other words, mean that researchers have less time for research and that they work less efficiently, thus causing other types of costs, as in the case of monitoring and performance-based pay.

Departments and faculties are also forced to compete.[16] Competition for research funding cannot, of course, be avoided, but if there is a shift towards external funds to be competed for instead of regular funding, too much time has to be spent on applying for funding, writing research outlines and annual reports and filling in forms rather than doing the research that is to be funded. It would also shift the power from research community to the sponsors and hence jeopardise academic freedom.

There is, in other words, no lack of competition within the university sector, but the competition is destructive. It diverts energy from the research and teaching (which in our experience requires as much concentration as research), and causes costs that are in principle measurable. A university would gain from more cooperation between individuals and departments, but cooperation does not flourish in the present climate.

Competition is just one example of how bureaucracy can emerge for unexpected reasons. Many authors have pointed out how the public management paradoxically tends to create bureaucracy. In some countries, for example in Finland, the government and some university bureaucrats also try to require each member of the staff to report how every half of an hour is spent. This can increase the output of a university only if there is slack, but otherwise it creates an additional cost in terms of lost research time (and would in addition be inconsistent with academic freedom). The information value of the reports that are created would probably be the same as under Soviet-style planning. Such ambitions are partly explained by lobbying from external sponsors. This provides another illustration of the dangers of handing over the main part of the responsibility for the research agenda to external bodies.

While we have pointed out that the new public management indirectly causes more of the bureaucracy that its rhetoric deplores, it is also important to emphasise that some administrative assignments are also inevitable in an ideal university. Participatory democracy would imply boards at different levels consisting of members of the academic staff. Such boards would make hiring decisions, and allocate budgets between departments and faculties. But the present top-down management misuses the labour time of the academic staff by its permanent revolution in the form of a never-ending sequence of reforms which either forces the boards to spend time on endless meetings, insofar as they do not imply transferring all power to professional managers.[17]

7. TOWARDS A DYNAMIC AND CREATIVE UNIVERSITY

The new public management has not yet completely transformed the university system. There is still in most cases no boss that would allocate daily assignments to the academic staff. Also, the academic freedom that still exists in most countries would in most universities rule out firing those who do a decent job but criticise the leadership, contrary to the rules of the business community (Hodgkinson, 2006). There is usually some form of representative democracy in Europe. But there are worrying tendencies towards a more managerial leadership, output funding and individual wage-setting.

The new public management greatly exaggerates the efficiency of the private sector which has been associated with excessive payments to managers and scandals such as Enron (Frey and Benz, 2005). We have attempted to provide a less one-sided picture of liberalisation and private and public activities elsewhere (see Willner, 2003; Björkroth et al., 2006; Willner and Parker, 2007). It has been pointed out that the present emphasis on business enterprises as role models is paradoxical, given recent stockmarket crashes, the excessive salaries and option arrangements for the managements (which are completely unrelated to their true performance).

This does not mean that a university has nothing to learn from the business community. According to W. Edwards Deming, there exist four ways to destroy a company by destroying work motivation and the pleasure of working. The first is to introduce competition between individuals and departments. The second is to introduce rewards and punishments related to variables that are difficult to measure. The third is to humiliate the employees, and the forth is to cause fear (Deming, 1993).

Universities should promote creativity, which is better encouraged by other instruments than rewards and punishments, such as dialogues within the research group, and the active participation of the project coordinators, who should not be just managers or accountants (Philipson, 1990). Also, role models are important: younger researchers learn from the enthusiasm displayed by their established colleagues (Lagerqvist, 1990).

There are therefore rational reasons for the following recommendations. A university does not work properly without proper funding. Salaries should be fixed and secure, so as to support intrinsic motivation, without excessive bureaucracy. The educational targets should be realistic. The university should become more autonomous, and the academic freedom should not be abolished but increased. Participatory democracy requires both an efficient senate and a representative congregation with the power to veto bad ideas from the top. Reforms of the framework in which research and teaching takes place should be cautious. The dynamism should be seen in offices, lecture halls and publications, not in administrative buildings and meeting rooms. Political decisions on universities should be informed, and destructive competition and the never-ending questioning of the right to exist for different university departments should stop.

Creativity should be encouraged by a climate of discussion rather than by rewards and punishments.

This may come across as a narrowly academic view on how academia should be managed. Let us therefore end by explaining how the business community can learn many lessons from traditional public management, including the governance of traditional universities. Frey and Benz (2005) mention fair and thorough hiring rules, career paths that discourage materialists, autonomous[18] employees within a set of given rules, and fixed-time contracts and restricted re-election opportunities for the senior management. In particular, they emphasise the intrinsic motivation, which should be supported by non-pecuniary rewards and fixed salaries, so as to avoid conflicts caused by difficulties of measurement and monitoring. Wage equality within each category would in addition release energy for common objectives, as a contrast to regimes of performance-related pay which encourage competition and cheating.

ACKNOWLEDGEMENT

This contribution is part of the Academy of Finland project *Reforming Markets and Organisations* (115003). Sonja Grönblom is grateful to The Research Institute for the Åbo Akademi Foundation, The Lars and Ernst Krogius foundation, The Foundation of Dagmar and Ferdinand Jacobsson, Otto A. Malm's donation fund and The Victoria foundation, for financial support.

NOTES

1. Hood (2001) even questions its usefulness, in particular because many countries have moved beyond it.
2. Such a strong leadership also requires longer terms in office for deans and vice-chancellors, because, as Schimank (2005) has observed, short periods means that leaders who make unpopular decisions would make enemies among their potential successors.
3. This might also explain the difficulty for a modern and formally more powerful vice-chancellor or dean to override a majority of professors (Schimank, 2005).
4. For example, the Congregation of the University of Oxford consists of 3552 representatives of its colleges and can veto any decision by the smaller university council which decides on most major issues, at the initiative of a number of members. The attempt to block a recent initiative to introduce a more managerial leadership is a case in point (Taylor, 2005).
5. Those who are hostile to the traditional university seem to think that it encourages research and teaching that is either irrelevant or even subversive (see Donoghue, 2004).
6. Output funding or budgeting by result can also be a rhetoric that conceals a planning system. In Finland, universities still receive money in proportion to the number of diplomas that are targeted in a four-year plan. But most universities are taking this

rhetoric seriously and use output funding as an internal allocation mechanism, in the belief that this might lead to a higher measurable output, just in case the Ministry of Education starts to punish those universities who produce less than planned.

7. For applications on involuntary unemployment, see for example Eaton and White (1983) and Shapiro and Stiglitz (1984).
8. This might be the case in particular under the perception that the service level would otherwise fall. A private employer would lose profits if an employee shirks and has therefore an incentive to adjust other inputs. A public employer can commit not to do that, so the employee knows that shirking would reduce the organisation's output (Francois, 2000).
9. The intrinsic motivation can also be misused in another way than by providing a low wage. It can cause the employer to invest less in other inputs than labour if they are substitutes to some extent (Glazer, 2004). The organisation would therefore work more efficiently if it was possible to ensure, for example, an adequate input of computer equipment or library resources.
10. Bénabou and Tirole (2003) cite the episode where Tom Sawyer persuades his friends to paint a fence by making the task seem so prestigious that they pay him for being permitted to paint. But self-realisation represents a different level in Maslow's hierarchy of needs (see Maslow, 1943); we are not saying that the academic staff should be manipulated so as to work for free.
11. The agent's effort supply curve might also be S-shaped: small rewards increase effort because they are not perceived as controlling, whereas somewhat higher rewards have the opposite effect because of motivation crowding-out (James, 2005). When the rewards are large enough, they may however again increase efforts, as in a standard model without intrinsic motivation.
12. The cognitive evaluation theory (CET) explains intrinsic motivation through psychological needs for autonomy and competence, so the impact of a reward depends on how these are perceived to be affected (Deci et al., 1999).
13. The cuts that have affected the universities in most countries may have contributed to a loss of self-esteem among the academic staff.
14. Those who pursue a business career are nowadays taught that lunches are only for losers and that friendship on the workplace is a bad idea (Hodgkinson, 2006), presumably because friendship distorts competition.
15. This is important, in particular, in Finland. To refer to budget cuts from a Finnish horizon may seem unjustified, because of Finland's comparatively high total R&D expenditures. However, the universities' share of these is only 19 per cent (Research in Finland, 2007).
16. For example, the State University of New York has adopted a Resource Allocation Method (RAM) that requires each of its campuses to be profitable, i.e to pay its own costs and generate its own revenues. Reduced costs are equated with increased quality in a way that has made critics to use the phrase 'Darwinism gone ape' (Scheuerman and Kriger, 2004).
17. Senate and council meetings can, for example, not be avoided, but it is essential that the academics set the agenda.
18. i.e. the right to make decisions independently within clear certain limits.

REFERENCES

Bénabou, R. and Jean Tirole (2002), 'Self-confidence and personal motivation', *Quarterly Journal of Economics*, **CXVII** (3), 871–915.

Bénabou, R. and J. Tirole (2003), 'Intrinsic and extrinsic motivation', *Review of Economic Studies*, **70** (3), no. 244, 489–520.

Besley, Timothy and Maitreesh Ghatak (2005), 'Competition and incentives with motivated agents', *American Economic Review*, **95** (3), 616–36.

Bitzer, J., W. Schrettl and P. Schröder (2004), 'The economics of intrinsic motivation in open source software development', mimeo, Free University, Berlin.

Björkroth, Tom, Sonja Grönblom and Johan Willner (2006), 'Liberalisation of public utility sectors: theories and practice', in Patrizio Bianchi and Sandrine Labory (eds), *International Handbook of Industrial Policy*, Cheltenham, UK and Brookfield, USA: Edward Elgar, pp. 180–97.

Deci, Edward L., Richard Koestner and Richard M. Ryan (1999), 'A meta-analytic review of experiments examining the effects of extrinsic rewards on intrinsic motivation', *Psychological Bulletin*, **125** (6), 627–68.

Deming, W. Edwards (1993), 'Presidential address: systems must rely on cooperation to reach winning outcomes', summary, *Amstat News*, American Statistical Association 202, October.

Donoghue, Frank (2004), 'The uneasy relationship between business and the humanities', *American Academic*, **1** (1), 93–110.

Eaton, B. Curtis and William D. White (1983), 'The economy of high wages: an agency problem', *Economica*, **May**, (50), 175–82.

Elster, Jon (ed.) (1986), 'Introduction', in Jon Elster (ed.), *The Multiple Self*, Cambridge: Cambridge University Press and Norwegian University Press, pp. 1–34.

Fehr, E. and A. Falk (2002), 'Psychological foundations of incentives', *European Economic Review*, **46**, 687–724.

Fehr, E. and U. Fischbacher (2002), 'Why social preferences matter – the impact of non-selfish motives on competition, cooperation and incentives', *Economic Journal*, **112**, C1–C33.

Francois, Patrick (2000), 'Public service motivation', *Journal of Public Economics*, **78**, 275–99.

Frey, Bruno S. (1997), 'On the relationship between intrinsic and extrinsic work motivation', *International Journal of Industrial Organization*, **15** (4), 427–40.

Frey, Bruno S. and Matthias Benz (2005), 'Can private learn from public governance?', *Economic Journal*, **115** (507), F377–F396.

Gintis, Herbert (2006), *Game Theory Evolving. A Problem-Centered Introduction to Modeling Strategic Interaction*, Princeton, NJ: Princeton University Press.

Glazer, A. (2004) 'Motivating devoted workers', *International Journal of Industrial Organization*, **22** (3) (March), 427–40.

Grönblom, Sonja and Johan Willner (2008), 'Privatization and liberalisation: costs and benefits in the presence of wage bargaining', *Annals of Public and Cooperative Economics*, **79** (1), 133–60.

Gruening, Gernod (2001), 'Origin and theoretical basis of the new public management', *International Public Management Journal*, **4**, 1–25.

Guttman, J.M. and A. Schnytzer (1989), 'Strategic work interactions and the kibbutz–kolkhoz paradox', *Economic Journal*, **99** (397), 686–99.

Henriksson, Linnea (2004), 'Marknadsorientering av kommunal serviceproduktion i Finland och Sverige – tre möjliga förklaringar till val av marknadslösningar', 'Market orientation of municipal service production in Finland and Sweeden – three possible explanations to the choice of market solutions' Åbo, Finland: Meddelanden från Ekonomisk-statsvetenskapliga fakulteten vid Åbo Akademi, A: 539.

Hess, James D. (1983), *The Economics of Organization*, Amsterdam: North-Holland.

Hodgkinson, Tom (2006), 'The winner takes it all', *New Statesman*, 3 July, 60–61
Holmström, Bengt and Paul Milgrom (1991), 'Multitask principal–agent analyses: incentive contracts, asset ownership and job design', *Journal of Law, Economics and Organization*, **7**, 972–91.
Hood, Christopher (1995), 'The "new public management" in the 1980s: variations of a theme', *Accounting, Organizations and Society*, **20** (2/3), 93–109.
Hood, Christopher (2001), 'Public management, new', *International Encyclopedia of the Social and Behavioral Sciences*, Elsevier, 12553-12556, accessed at www.christopherhood.net/pdfs/npm_encyclopedia_entry.pdf.
Huisman, Jeroen, Harry de Boer and Leo Goedgebuure (2006), 'The perception of participation in executive governance in Dutch universities', *Tertiary Education Management*, **12**, 227–39.
Jacobs, Bas and Frederick van der Ploeg (2006), 'Guide to reform of higher education: a European perspective', *Economic Policy*, **47**, July, 357–92.
James. Harvey S. (2005), 'Why did you do that? An economic examination of the effect of extrinsic compensation on intrinsic motivation and performance', *Journal of Economic Psychology*, **26**, 549–66.
Jensen, M.C. and K.J. Murphy (1990), 'Performance pay and top-management incentives', *Journal of Political Economy*, **98**, 225–64.
Lagerqvist, Ulf (1990), 'Den undanglidande gnistan', in Georg Klein (ed.), *Om kreativitet and flow*, Värnamo, Sweden: Brombergs, pp. 26–39.
Lawton (2005), 'Public service ethics in a changing world', *Futures*, **37**, 231–43.
Lepage, Henri (1980), *I morgon kapitlism*, Avesta, Sweden: Ratio.
Maslow, Abraham (1943), 'A theory of human motivation', *Psychological Review*, **50** (4), 370–96.
McPherson, Michael S. and Morton Owen Shapiro (1999), 'Tenure issues in higher education', *Journal of Economic Perspectives*, **13** (1), 85–98.
Megginson, W.L. and J.M. Netter (2001), 'From state to market: a survey of empirical studies on privatization', *Journal of Economic Literature*, **XXXIX** (2), 321–89.
Minkler, L. (2004), 'Shirking and motivation in firms: survey evidence on worker attitudes', *International Journal of Industrial Organization*, **22** (6), 863–84.
Newberry, Susan and June Pallot (2004), 'Freedom or coercion? NPM incentives in New Zealand central government departments', *Management Accounting Research*, **15**, 247–66.
Philipson, Lennart (1990), 'Den kreativa gruppen', in Georg Klein (ed.), *Om kreativitet and flow*, Värnamo, Sweden: Brombergs, pp. 116–22.
Research in Finland (2007), The Ministry of Education, www.research.fi/en/muut/Research_in_Finland_2007.pdf, accessed 15, February 2009.
Schelling, Thomas C. (1978), 'Egonomics, or the art of self-management', papers and proceedings, *American Economic Review*, **69** (2), 290–94.
Scheuerman, William and Thomas Kriger (2004), 'Introduction – the concept of corporatization: a useful tool or a feel-good slogan?', *American Academic*, **1** (1), 7–19.
Schimank, Uwe (2005), '"New public management" and the academic profession: reflections on the German situation', *Minerva*, **43**, 361–76.
Shapiro, Carl and Joseph E. Stiglitz (1984), 'Equilibrium unemployment as a worker discipline device', *American Economic Review*, **74** (June), 433–44.
Sveiby, Karl Erik and Anders Risling (1987), *Tietoyrityksen johtaminen – vuosisadan haaste?*, Esbo, Finland: Weilin & Göös.

Taylor, Matthew (2005), 'Dons fear betrayal as Oxford looks to business', *The Guardian*, 30 April, accessed at http://education.guardian.co.uk/administration/story/0,9860,1473668,00.html.

Willner, Johan (2001), 'Ownership, efficiency, and political interference', *European Journal of Political Economy*, **17** (4), 723–48.

Willner, Johan (2003), 'Privatization – a sceptical analysis', in David Parker and David Saal (eds), *International Handbook of Privatization*, Cheltenham, UK and Northampton, MA, USA: Edward Elgar, pp. 60–86.

Willner, Johan and Sonja Grönblom (2009), 'Do budget cuts and incentive wages lead to higher productivity in a university?', *International Review of Applied Economics*, (forthcoming).

Willner, Johan and David Parker (2007), 'The performance of public and private enterprise under conditions of active and passive ownership and competition and monopoly', *Journal of Economics*, **90** (3), 221–53.

Winston, Gordon C. (1999), 'Subsidies, hierarchy and peers: the awkward economics of higher education', *Journal of Economic Perspectives*, **13** (1), 13–36.

7. A theoretical analysis of the relationship between social capital and corporate social responsibility: concepts and definitions

Lorenzo Sacconi and Giacomo Degli Antoni

1. INTRODUCTION

Trust, trustworthiness and ethical norms of reciprocity and cooperation have been receiving more and more attention in economic analysis. In particular, two concepts have been widely used in order to study the socio-economic effects of these factors: the concept of social capital (hereafter SC) and of corporate social responsibility (hereafter CSR).

After the seminal work by Putnam et al. (1993) that revealed the effect of SC on economic and government performance, many definitions of social capital have been introduced in the literature and have been considered in order to analyse the role of interpersonal relations in affecting economic activity by favouring cooperation.[1]

Different approaches characterize also the notion of CSR. If we look at the stakeholder approach (Freeman, 1984, 2000; Freeman and Evan, 1990) or at the contractarian approach to CSR (Sacconi, 2004, 2006, 2007 a,b), relational aspects, in terms of trust, trustworthiness and spirit of cooperation, may have a key role in promoting the coordination processes between firm and stakeholders that are essential in order to implement the CSR practices.[2]

Even though SC and CSR seem to be linked by many common elements related to the quality and quantity of social relations between agents, their relationship has not been deeply investigated yet. This chapter is aimed at shedding light on some aspects of this relationship, in particular, by investigating the idea of a virtuous circle, between the level of SC and the implementation of CSR practices, that fosters socio-economic development by generating social inclusion and social networks based on trust and trustworthiness.

Following the literature on SC that stresses its multidimensional character (e.g. Paldam, 2000), we consider two dimensions of this notion. Starting from the distinction introduced by Uphoff (1999), we take into account a cognitive and a structural idea of SC. The first one essentially refers to the dispositional characters of agents that affect their propensity to behave in different ways. The latter refers to social networks connecting agents.

With regard to the concept of CSR, we adopt a contractarian approach and consider CSR as an extended model of corporate governance, based on the fiduciary duties owed to all the firm's stakeholders (Sacconi, 2006, 2007a). Among stakeholders, we distinguish between 'strong' and 'weak' stakeholders. Both these two categories have made specific investments in the firm. However, strong stakeholders are precious for the firm because they bring in strategic assets. They are, for example, skilled workers or institutional investors. On the other hand, weak stakeholders do not bring strategic assets into the firm and firms have material incentives at defecting in the relationship with them. They are, for example, unskilled workers.

Considering the notions of cognitive and structural SC and a contractarian approach to CSR, we show that:

1. The level of cognitive SC plays a key role in inducing the firm to adopt and observe CSR practices that respect all the stakeholders;
2. The decision of adopting formal instruments of CSR contributes to create cognitive SC that is endogenously determined in the model;
3. The level of cognitive SC and the decision to adopt CSR practices creates structural SC in terms of a long-term relationship between the firm and the weak and strong stakeholders.

This chapter contributes to the literature on social capital, CSR and social networks in two ways. First we take seriously the problem of defining social capital and analyse the theoretical relationship between two specific forms of SC. We distinguish between a cognitive and a structural dimension of SC and show under which condition cognitive SC can contribute to the creation of social networks.

Second, the chapter examines the complementarity between SC and CSR, showing that they generate a virtuous circle that creates favourable conditions for socio-economic development. We are not aware of previous studies on this specific topic.

The chapter is divided into five sections. In the second and third sections we define respectively the concept of SC and CSR adopted in the chapter. In section 4 we extensively discuss the theoretical connections between the definitions of SC and of CSR. Concluding remarks follow.

2. A MULTIDIMENSIONAL APPROACH TO SOCIAL CAPITAL

Starting from the contributions by Coleman (1988, 1990) and Putnam et al. (1993), many definitions of SC have been proposed and an agreement on a commonly accepted definition has not been reached. Nevertheless, the multidimensional nature of SC seems to be commonly recognized. We can distinguish at least between two main characterizations of this notion. On one hand, there are definitions that mostly look at the networks that constitute the structure of relations of a single agent or of a community as a whole (Coleman, 1988; Burt, 1992, 2002; Lin, 2001). On the other hand, SC is defined by looking mostly at cultural and mental factors, such as attitude and norms (Putnam et al., 1993; Knack and Keefer, 1997). Even though these two approaches are linked, there are few analytical studies on the relationship of cause and effect between the definitions of SC in terms of social networks and of attitude and norms. In this perspective, one of the attempts is by Uphoff (1999) who distinguishes and analyses the interrelations between two categories of social capital: structural SC and cognitive SC. According to Uphoff's definition 'The structural category is associated with various forms of social organization, particularly *roles*, *rules*, *precedents* and *procedures* as well as a variety of *networks* that contribute to cooperation, and specifically to mutually beneficial collective action (MBCA), which is the stream of benefits that results from social capital. The cognitive category derives from mental processes and resulting ideas, reinforced by culture and ideology, specifically norms, values, attitudes, and beliefs that contribute cooperative behavior and MBCA' (Uphoff, 1999, p. 218). In particular, Uphoff stresses that networks 'are crucially sustained by expectations (that is, by norms) of reciprocity' and that this reveals the existence of an essential cognitive dimension of networks. Starting from the classification proposed by Uphoff, we define structural SC as cooperative networks between agents, and focus on two specific elements of cognitive SC: beliefs and dispositions. Beliefs and dispositions are different to norms and values in that they have a micro-dimension because they are referred to as single agents. Dispositions can be affected both by macro variables (norms and values shared by the community where agents live) and by micro elements (genetic and psychological factors). Beliefs depend essentially on past experience and on mutual agreement concerning the respect of specific commitments.

Table 7.1 shows the main features of the two categories of SC according to our definition.

Beliefs and dispositions are the constitutive elements of our notion of

Table 7.1 Categories of social capital

	Cognitive		Structural
Constitutive elements	Beliefs	Dispositions	Cooperative linkages
Determinants	**Ex ante** Agreements Ethical commitments Framing **Ex post** Learning from other previous behaviour Past experience	Shared norms Values Genetic and psychological factors	Beliefs that other will be cooperative Disposition to cooperate Endogenous social sanctions against defections
Effects	Trust Cooperation	Trust Trustworthiness Cooperation	Cooperative relations

Source: Authors.

cognitive SC. Beliefs in the behaviour of others depend on the behaviour they have already shown in the past. Moreover, beliefs can be generated or reinforced by ethical commitments that agents take (e.g. subscribing to an agreement on a ideal principle). Dispositions principally stem from the norms and values shared in the community where the agents grow up, but they also depend on micro elements such as genetic and psychological factors. Both beliefs and dispositions can promote (or, obviously, reduce) trust and propensity to cooperate.

Structural social capital is constituted by cooperative linkages between agents. We consider essentially three factors that can promote the creation of cooperative relations: (1) beliefs that others will be cooperative, (2) disposition to cooperate and (3) the existence of effective sanctions that punish the agents that do not cooperate.

After having introduced in the next section the approach to CSR adopted in this chapter, we will refer to the distinction between structural and cognitive SC in order to analyse the relationship between SC and CSR. In particular, our aim is to show how social capital and CSR interact generating a virtuous circle that increases the initial endowment of both these factors in a society.

3. THE APPROACHES TO THE CONCEPT OF CORPORATE SOCIAL RESPONSIBILITY

There are essentially three different ways to approach the concept of CSR. The first one is to deny or limit substantially the room for the social responsibility of firms. According to Friedman (1977), the only social responsibility of a firm is to make profits respecting the rules, that means without breaking the law. This idea is founded on two arguments. The first one concerns the principle of the maximization of the shareholder's value. A manager must run a firm pursuing the interests of the group of people that s/he represents because they have given him/her the control of the firm. The second one comes from the firm belief that the maximization of the shareholder's value implies an optimal use of resources and, consequently, the maximization of total wealth. Jensen (2001) supports the idea that a firm should first pursue the shareholder's value maximization. This author stresses that the substitution of this aim with another one – e.g. the maximization of a function that explicitly includes the utility of all the stakeholders – could introduce the inability to provide a clear benchmark against which management strategies and company performances can be assessed, this indeterminacy also being the basis for the charge of opening the route to opportunistic behaviour on the part of managers. Moreover, Jensen says that the maximization of shareholder's value is able, in the long run, to solve problems and also take into account the interests of stakeholders that the stakeholder approach to CSR wants to satisfy.

A second way to look at the CSR is to interpret the decision of adopting practices aimed at considering interests of subjects other than the shareholders as a kind of philanthropy (Baron, 2005).

Finally, according to a more articulated approach to CSR, whoever runs a firm has to take into account the interests of all the stakeholders. Within this approach it is possible to distinguish between the stakeholder approach and the contractarian approach. The stakeholder approach was introduced by Freeman (1984) who stressed the idea of ethical balance between the interests of the firm and stakeholders in a perspective of strategic management. The contractarian approach to CSR differs from the stakeholder approach principally because it aims at specifying, through a rational agreement (i.e. the social contract), a criterion for defining a balance of the firm's stakeholders' interests.

In this chapter we adopt the contractarian approach in order to study the relationship between SC and CSR for two main reasons. First, because the contractarian approach makes it possible to run a firm according to a multi-stakeholder approach by introducing a criterion for defining a balance among the firm's stakeholders. Second, because the social contract underlies the relations between our notions of SC and CSR. In fact, as we will clarify in section 4, it is the social contract which allows:

1. The activation of the agent's beliefs on the firm's behaviour which are a constitutive element of our notion of cognitive SC;
2. The firm to develop a reputation and to induce its stakeholders to start cooperative and fiduciary relations which represent our concept of structural SC.

3.1 The Contractarian Approach to CSR

The contractarian approach to CSR stems from the idea that a firm is an institution that arises in order to solve the incompleteness of contracts and bounded rationality. In a context characterized by incompleteness of contracts and bounded rationality, economic institutions allocate through property rights and hierarchical organizations, decision rights to certain parties in any subset of the economy. The need for general and abstract ethical principles rises from the risk that this discretion may be abused.

Within the theoretical framework of the contractarian approach, we define CSR as a 'model of extended corporate governance whereby those who run a firm (entrepreneurs, directors and managers) have responsibilities that range from fulfilment of their fiduciary duties towards the owners to fulfilment of analogous fiduciary duties towards all the firm's stakeholders' (Sacconi, 2006, p. 262).

In order to clarify the introduced definition of CSR, two notions require to be expanded: the concept of fiduciary duty and of stakeholders.

The notion of fiduciary duties, refers to a situation where a subject has a legitimate interest but is unable to make the relevant decisions in the sense that s/he does not know what aims to pursue, what alternative to choose, or how to deploy his/her resources in order to satisfy his/her interest. This subject, the *trustor*, can delegate decisions to a *trustee* giving him the power to choose actions and goals. The trustee may thus count on the resources of trustor and select the appropriate course of action. For a fiduciary relationship to arise, the trustor must have a claim (right) towards the trustee. In other words, the trustee acts and uses the resources made over to him/her in order to achieve results that satisfy (to the best extent possible) the trustor's interests. These claims (i.e. the trustor's *rights*) impose fiduciary duties on the agent who is entitled with authority (the trustee), which s/he is obliged to fulfil.[3]

By the term 'fiduciary duty', therefore, we mean the duty (or responsibility) to exercise authority for the good of those who have granted that authority and are therefore subject to it.

The term 'stakeholders' denotes individuals or groups with a major stake in the running of the firm and who are able to influence it significantly (Freeman and McVea, 2002). Different categories of stakeholders can be specified. We introduce an original distinction between strong and weak stakeholders. Both these categories make specific investments in the firm.[4] The key element that allows the distinction between strong and weak stakeholders concerns the consequences that the break in the relationship with the firm produces both on the stakeholder and on the firm.

Strong stakeholder The difference between the discounted payoff that strong stakeholders and firms get cooperating forever and defecting at the first stage (and not cooperating ever again) is positive. Strong stakeholders bring strategic assets into the firm. They are for example institutional investors or highly skilled workers.

Weak stakeholder Weak stakeholders would like to cooperate forever with the firm, but the discounted payoff that the firm gets cooperating forever with them is lower than the payoff it obtains defecting at the first stage and not cooperating ever again. Weak stakeholders do not bring strategic assets into the firm. They are for example ordinary investors, unskilled workers or unskilled contractors.

Note that the firm (or the stakeholder) defects in the relationship with the stakeholder (or the firm) when it behaves opportunistically by trying to obtain all the surplus generated through the interaction. This behaviour concerns a long-term relationship between the firm and its stakeholders and it does not necessarily imply the immediate interruption of the relation. In fact, defection does not coincide with the break in the relationship but makes the counterpart indifferent in continuing or stopping it. For this reason, obviously, the defection increases the probability that the subject who suffers from the opportunistic behaviour breaks the relationship.

When the relation concerns a firm and a strong stakeholder both have the economic incentive to avoid opportunistic behaviour because the payoff obtained by defecting does not compensate the loss connected to the break in the relationship which could follow the defection. On the contrary, in the relationship between a firm and a weak stakeholder, the former does not fear the break in the relationship and there are not endogenous economic incentives for the firm to avoid the defection.

The definition of CSR as an extended responsibility towards its stakeholders is rooted in neo-institutional theory (Williamson, 1975, 1986;

Grossman and Hart, 1986; Hart and Moore, 1990; Hart, 1995; Hansmann, 1996). According to this theoretical approach, the firm emerges as an institutional form of 'unified transactions governance' aimed at remedying imperfections in the contracts that regulate exchange relations among subjects endowed with diverse assets. The joint use of these assets generates a surplus. Agents must find an agreement on the conditions characterizing their exchange relations. The agreements essentially concern: the reciprocal investment that must be realized to generate the surplus and the way of dividing the surplus. The contracts that have to regulate these agreements cannot be complete. They do not include provisos referred to unforeseen events, either because of the costs of drafting them, or because the cognitive limits of the human mind that make it impossible to predict all possible states of the world. It generates the risk of opportunistic behaviour in case the unforeseen events happen and the renegotiation of the contracts' conditions becomes a necessity. For fear of opportunistic behaviour that the party in a stronger *ex post* position can have appropriating the entire surplus, thereby expropriating the other stakeholders, agents, that expect to be expropriated, will have no incentive to undertake their investments at the optimal level. This expectation of unfair treatment can generate a loss of efficiency at the social level.

The firm responds to this problem by bringing the various transactions under the control of a hierarchical authority – the authority, that is, of the party which owns the firm and through ownership is entitled to make decisions over the contingencies that were not ex ante contractible.[5] This party is thus safeguarded against opportunism by the other stakeholders and will realize the optimal level of investment. Nevertheless, there is a risk of abuse of authority against the other parties (Sacconi, 1999, 2000, 2006). Those wielding authority, in fact, are able to threaten the other stakeholders. The former can exclude the latter from access to physical assets of the firm, or from the benefits of the contract, to the point that those other stakeholders become indifferent between the decision to accept the expropriation and the decision to forgo the value of their investments by withdrawing from the relationship. Thus the entire surplus, included that part produced by the efforts and investments made by the non-controlling stakeholders, will be appropriated by those in a position of authority. Again forward-looking non-controlling stakeholders will be deterred from entering a relationship with the controlling party.

Many stakeholders among the non-controlling ones, will *ex ante* be discouraged from investing (if they foresee the risk of abuse), while *ex post* they will resort to conflicting or disloyal behaviour (typically possible when asymmetry of information is inherent in the execution of some subordinate activity) in the belief that they are being subjected to abuse of authority. Therefore, the optimal level of investment could not be achieved and a second-best solution arises. All governance solutions based on the allocation of property rights to a single party may approximate social efficiency, but they can never fully achieve it.

The relative (in)efficiency is due to manifest or simply expected unfairness. This is the reason why non-optimal level of investment are realized by non-controlling stakeholders and it underlines the role that fairness plays in affecting efficiency, at least with regard to the real-life problem of working out an acceptable solution for the governance of transactions.

According to the contractarian approach adopted in this chapter, when CSR is viewed as 'extended governance', it can complete the firm as an institution of transactions governance (Sacconi, 2000, 2006). The firm's legitimacy deficit is remedied if the residual control right is associated with further fiduciary duties towards the subjects that face the risk of abuse of authority and are deprived of that residual control right. At the same time, this generates an increase in social efficiency because it reduces the disincentives and social costs generated by the abuse of authority. In order to avoid the second-best solution due to the risk of abuse of authority, the firm must be grounded on a rational agreement (the constitutional contract of the firm) between those who run the firm and the non-controlling stakeholders (see Sacconi, 2006). The constitutional contract of the firm is the basis not only for the allocation of control over the firm but also to include in this structure, other rights – essentially responsibility claims in defence of stakeholders other than those protected by the property right. The resulting institutional structure defines the principles of the firm's governance structure consistently with the notion of CSR as a governance model with multiple fiduciary duties.

From this perspective, 'extended governance' should comprise:

- The residual control rights (ownership) allocated to the stakeholder with the largest investments at risk and with relatively low governance costs (as well as the right to delegate authority to professional directors and management);
- The fiduciary duties of those who effectively run the firm (administrators and managers) towards the owners, given that these have delegated control to them;
- The fiduciary duties of those in a position of authority (the owner or the managers) towards the non-controlling stakeholder. In particular considering the obligation to run the firm so that the non-controlling stakeholders are not deprived of their fair shares of the surplus produced from their specific investments, and that they are not subject to negative externalities.

The constitutional contract of the firm provides that authority should be delegated to the stakeholder most efficient in performing governance functions and defines the fiduciary duties of this part towards the non-controlling stakeholders.[6] In fact, the stakeholders agree to submit to authority, thereby rendering it effective, only if the contract contains the proviso that stipulates that the firm's new governance structure must comply with fiduciary duties towards all the stakeholders (owners and non-owners). Otherwise, the risk of abuse of authority cannot be overcome.

The definition of social contract and constitution of the firm stems from the solution of a bargaining cooperative game in which stakeholders must agree on a shared action plan (a joint strategy) which allocates tasks among the members of the team so that the contribution of each of them is efficient (because it produces the maximum surplus net of each stakeholder's costs) and defines the sharing of the surplus that is generated by the cooperation.[7]

After having specified the contractarian approach to the CSR that we adopt in this chapter, after having argued its theoretical foundation in the framework of the neo-institutional theory, and after having introduced the role of the social contract in facing the risk of abuse of authority, we have to analyse the implementation aspects of this notion of CSR. In particular, we are going to investigate if the social contract is also able to induce endogenous incentives and motivations for the firm to adopt the normative model of extended fiduciary duties. Moreover, we will show how this model can be implemented by the firm.

The idea is that the incentives related to the formation of reputation can play a key role in the firm's decision to endorse and respect extended fiduciary duties towards all the stakeholders. The stakeholders will decide to cooperate with the firm if they trust that it will not abuse them: if stakeholders observe that the firm always respects the social contract, it will increase its reputation and stakeholders will decide to invest at an optimal level into the firm. If they observe opportunistic behaviour by the firm, its reputation will dramatically diminish.

The problem with regard to the creation of reputation arises because the relations between the firm and its stakeholders are characterized by settings in which information or knowledge about the action of the firm is incomplete or highly asymmetric. Because of incomplete information, the stakeholders cannot verify if the firm has actually behaved as a honest cooperative agent by trying to avoid any opportunistic behaviour. Incomplete information essentially eliminates the possibility for the firm to develop a reputation.

In order to avoid the consequences caused by incomplete information on the formation of reputation, the firm must subscribe to an explicitly announced standard that sets out general principles, whose contents are such to elicit stakeholder consensus, as well as explicit commitments to compliance with principles and rules which are to be known *ex ante* by stakeholders. The standard must contain explicit norms with an appropriate structure that must be endorsed by the firm and established in the light of a multi-stakeholder social dialogue, such as to induce impartial acceptability.

It is the standard that enables the social mechanism of reputation to function properly by allowing stakeholders to increase their trust in the firm and in its compliance with CSR principles. The standard and the procedures ensuring compliance with it are announced *ex ante*; and it is *on these* – not in relation to particular (unforeseen) events or to particular (unobservable) actions or outcomes – that the firm and stakeholders pass homogeneous judgement on *ex post* compliance with them. The reputation is created if the behaviour of the firm is coherent with the principles declared in the standard.[8] Compliance with CSR voluntary but explicit norms (codes of ethics, management system standards, etc.) can solve the incomplete information problem and can allow the firm to develop its reputation and get its share of surplus produced through the cooperation with the stakeholders.

4. SOCIAL CAPITAL AND CORPORATE SOCIAL RESPONSIBILITY: A THEORETICAL ANALYSIS

In this section we analyse the relationships between SC and CSR. Our principal aim is to show how SC and CSR interact generating a virtuous circle that increases the initial endowment of both these factors in a society. We identify three main levels of interaction between social capital and CSR.

1. In the first, cognitive SC, understood as dispositions, is an input of CSR. The more individuals who are in contact with firms have disposition to cooperate with agents who respect principles of cooperation, the more firms have incentives to develop a reputation by adopting a CSR standard that declares their compliance with the principles.
2. In the second, the adoption of an explicit CSR standard generates cognitive SC, in terms of beliefs. The commitments to compliance with CSR principles, in terms of fulfilment of fiduciary duties towards all the stakeholders, contributes to determine the beliefs of stakeholders on the cooperative behaviour of firms.

3. Finally, cognitive SC, both dispositions and beliefs, and CSR create the economic incentives that induce the firm to completely fulfil its commitments towards all its stakeholders. It means that firms will decide to cooperate with all their stakeholders creating a cooperative network that would not be created in the absence of cognitive SC and of the adoption of CSR standards.

4.1 Conformist Preferences and Cognitive Social Capital

In order to attain this aim, first we have to specify better the concept of cognitive social capital in terms of belief and disposition. We start from the assumption that agents have motives to act that are not purely geared to material advantages (consequences of actions), but which extend beyond mere material advantage. In particular, we assume that agents have conformist preferences (Grimalda and Sacconi, 2002, 2005; Sacconi, 2007a) that are defined over states of affairs that are described as sets of interdependent actions characterized in terms of their degree of conformity to a given abstract principle or ideal.

The utility function of agents that have motives to act that depend also on conformist preferences is:

$$V_i = U_i(\sigma) + \lambda_i F[T(\sigma)].$$

The first term $U_i(\sigma)$ is the material utility got by agent i in state σ. The second term is the ideal utility and reflects the agent's concern with other types of reasons to action, meant in general as the degree of conformity of the social state of affairs (σ) – the agent's and the other participants' behaviours – to the normative principle of welfare distribution T.

λ_i is an exogenous parameter that represents the disposition to conform to the ideal principle T given the beliefs in the others' behaviour. The motivation to conform to the principle T for agent i depends on the value of λ_i. The higher λ_i is, the more the agent i will be disposed to conform to the principle T if he/she believes that the others will conform to the principle. λ_i represents the endowment of cognitive social capital (in terms of disposition) of agent i.

The effects on ideal utility of beliefs (in the degree of conformity to the ideal of other agents) is captured by the function F. Following Grimalda and Sacconi (2002), we adopt a particular specification for F based on an idea of expected mutuality in conforming to the normative prescriptions. If we consider a two-person game, F can be specified by considering two elements:[9]

1. f_i: the index of conditional conformity of player i. The value of this index depends on how much the player i contributes to carry out the ideal T with his/her behaviour (i.e by conforming or deviating from the ideal T), given what s/he believes about the other player's choice.
2. $\tilde{f}_j$: the esteem that player i forms about j's compliance with the ideology. The value of this index depends on how much the other player contributes to carry out the ideal T with his/her behaviour (i.e by conforming or deviating from the ideal T), given what second player believes (and first player believes that second player believes) that first player will do.

These two indices contribute to determine F and the utility function becomes:

$$V_i(\sigma_i, b_i^1, b_i^2) = U_i(\sigma_i, b_i^1) + \lambda_i[1 + \bar{f}_j(b_i^1, b_i^2)][1 + f_i(\sigma_i, b_i^1)]$$

Where b_i^1 is the first order belief that player 1 has in the action of player j. b_i^2 is the second order belief about player j's belief in the action adopted by player i.

Both beliefs and dispositions play a key role in determining the (ideal) utility of the stakeholder i:

1. If i conforms totally to the ideal principle T and believes that j will conform totally to the ideal, then the ideal utility of i will assume the maximum value:

$$\lambda_i \times 1 \times 1 = \lambda_i$$

2. If i does not conform completely and believe that also j will not conform completely, the value of ideal utility will be lower than λ_i:

$$(1 - x)(1 - y)\lambda_i < \lambda_i$$

3. Finally, if the conformity of one of the two agents is zero, the ideal utility got by agent i goes to zero:

$$(1 - 1)(1 - y)\lambda_i = 0$$

The belief in the behaviour of others with regard to their conformity with the ideal principle T, and the disposition, to conform to T given the belief, determine the value of the ideal utility got by agents that have conformist preferences and are the two elements that constitute our notion of cognitive social capital. The disposition λ is generated both by micro and macro factors. First it is related to psychological and genetic factors that affect the disposition of each individual. Second, it is affected by the culture and social norms that characterize the community where the agents live. Belief of agents in the degree of conformity to the principle of others depends on two factors strictly interrelated. First, beliefs can arise in relation to a rational agreement that agents subscribe where they declare the decision to respect and conform to the principle T. Second, they depend on the past behaviour of others (that can confirm or not their actual willingness to conform).

4.2 The CSR Principle and the Relationship between CSR and Cognitive Social Capital

After having clarified the definition of cognitive SC in terms of dispositions and beliefs, we introduce the notion of CSR and investigate the connection between CSR and cognitive SC with regard to two specific classes of agents: the firm and its stakeholders. The definition of CSR as an extended model of corporate governance, based on the fiduciary duties owed to all the firm's stakeholders enters in this analysis essentially in relation to the principle T. In our analysis T is the abstract and general principle that the firm must explicitly endorse if it wants to develop a reputation that can induce stakeholders to enter in cooperative relations with it. As we showed in section 3.1, if the firm wants to be trusted by stakeholders, it must subscribe an explicitly announced standard that sets out general principles. This standard allows stakeholders to increase their trust in the firm and in its compliance with the (CSR) principles. A characterisation in contractarian terms of the ideal principle T is given by the Nash bargaining solution, called also Nash social welfare function N:

$$T(\sigma) = N(U_1, \ldots, U_n) = \prod_{i=1}^{n} (U_i - d_i)$$

where d_i stands for the reservation utility that agent i can obtain when the bargaining process collapses.

The adoption of the CSR principle by the firm is also interconnected with both the beliefs and the dispositions of the stakeholders who have to decide whether to enter into a cooperative relation with the firm or not.

First, the disposition λ of stakeholders may give the firm the incentive to adopt the CSR standard. In fact, according to our definition of conformist preferences, when λ_i increases, it increases also the utility of the stakeholder i in cooperating with a firm that conforms to the principle T. Because of that, stakeholders characterized by high levels of λ will decide to trust a firm that declares and respects CSR principles sooner than stakeholders with low levels of λ. It means that a firm, which acts in a context where the stakeholders are endowed by high cognitive SC, will be able to develop a reputation faster and have lower costs than a firm which is related to stakeholders who do not have high endowment of cognitive SC (in terms of λ). Obviously the firm is interested in developing a reputation because it induces stakeholders to invest their resources in the firm at an optimal level with positive effects on the firm's activity.

Second, the adoption of a CSR standard affects the beliefs of the stakeholders. It is only through the explicit declaration of the principle T that stakeholders can form their beliefs on the type of the firm they are related to. Without the adoption of a CSR standard, because of the incomplete information that characterizes the relations between the firm and its stakeholders (section 3.1), the stakeholders cannot form their belief in the conformity of the firm to the principle and, consequently, cannot obtain the ideal utility which strictly depends on beliefs.

Dispositions, beliefs and the adoption of a CSR standard can induce stakeholders to cooperate with the firm and to undertake their investments at the optimal level. The stakeholders get a positive material utility in cooperating with the firm until the firm will not abuse them. Moreover, the stakeholders get ideal utility until they observe that the firm conforms to the CSR principle, that is it fulfils its fiduciary duties towards all the stakeholders. If a stakeholder observes opportunistic behaviour by the firm, its reputation will dramatically diminish. A stakeholder loses his/her ideal utility in cooperating with the firm both if the firm abuses him/her and if the firm abuses another stakeholder and the loss of the ideal utility depends on the value of λ.

4.3 The Relationship between CSR, Cognitive Social Capital and Structural Social Capital

After having discussed the relationship between CSR and cognitive SC, we analyse the role that these two elements have in promoting the creation of structural SC in terms of cooperative relations between the firm and its stakeholders. According to the definitions we introduced in section 3.1, the stakeholders of a firm can be divided into two classes. Strong and weak stakeholders. Looking only at their monetary payoffs, both strong stakeholders and the firm are reciprocally interested in cooperating. In contrast, the cooperative relation between weak stakeholders and the firm is asymmetric and not sustainable in the long term. The firm has economic incentives in defecting in the relation with weak stakeholders who do not bring strategic assets into the firm.

Obviously, both strong and weak stakeholders are under the risk of abuse of authority. Nonetheless, the strong stakeholders have the real possibility of punishing the firm which abuses by stopping cooperation with it.[10] The gain that the firm gets by cooperating with the strong stakeholders reduces the risk of abuse for the latter. Conversely, the weak stakeholders do not have any possibility of avoiding the abuse, because the firm is not interested in starting a cooperation process with them and, consequently has no fear of a sanction by the weak stakeholders.

According to our idea, if (strong) stakeholders are endowed by high cognitive SC in terms of dispositions to cooperate with agents who conform to principles of cooperation and they believe that the firm will be cooperative with all the stakeholders, then the firm who contradicts these beliefs by behaving opportunistically with weak stakeholders faces the sanction of the strong ones who may decide to stop cooperating with it. For this reason, in presence of an appropriate structure of dispositions and beliefs, the punishment of strong stakeholders, that consists in stopping their cooperation with the firm, may be a protection against opportunistic behaviour of the firm also for the weak stakeholders.

Let us start from a situation in which the stakeholders of the firm (consumers, suppliers etc.) are characterized by a high level of λ. According to the argument developed in the previous section, this firm will have incentives to adopt the CSR standard. In fact, stakeholders with high level of λ will decide to trust a firm that declares and respects CSR principles sooner than stakeholders with low levels of λ. For this reason, a firm, who acts in a context where stakeholders are endowed by high cognitive SC, will be able to develop a reputation faster and suffer lower costs than a firm which is related to stakeholders who do not have high endowment of cognitive SC (in terms of λ). It is the adoption of the CSR standard which allows the stakeholders to form their beliefs in the respect of cooperative principles by comparing the firm's behaviour with its CSR declaration. The CSR standard, beliefs and dispositions generate a positive ideal utility that the stakeholders get by cooperating with the firm who conforms to the CSR principle.

If the firm decides to stop cooperating with weak stakeholders, because its material payoffs are higher if it defects than if it cooperates with them, then it stops conforming with the ideal CSR principle. If the strong stakeholders of the firm have conformist preferences and they conform to the CSR principle that concerns the fulfilment of the duties towards all the stakeholders, when the firm behaves opportunistically with the weak stakeholders, they lose their ideal utility. If λ, that is the weight of the ideal utility in the agents' utility function, is high enough, the strong stakeholders may decide to punish the firm which abuses the weak stakeholders. In particular, they may be ready to lose the investment they have made in the firm, i.e. the material utility that they get by cooperating with it, in order to start a new cooperative relation with a firm who conforms to the principle of cooperation. This represents a sanction for the firm that loses the gain that it gets by cooperating with strong stakeholders.

The possibility that the strong stakeholders stop their cooperation is a credible threat for the firm because it is connected with endogenous incentives that are determined by the role of cognitive SC on stakeholders' ideal utility. The fear of being punished by the strong stakeholders can induce the firm not to abuse the weak stakeholders, thus making cooperative relations sustainable also between the firm and its weak stakeholders.

Starting from a precise definition of cognitive SC, from a contractarian approach to CSR and from the assumption that agents have motives to act that are not purely geared to material advantages (conformist preferences) our theoretical argument leads to a positive relation between cognitive social capital, CSR practices and structural social capital. Cognitive SC and CSR, by reciprocally interacting, generate endogenous incentives for the firm to behave cooperatively with weak stakeholders. In this perspective, they generate the condition for the creation of structural social capital that would not be created otherwise.

Figure 7.1 summarizes the theoretical relations between cognitive SC, adoption of the CSR standard, creation of reputation and creation of cooperative relations between the firm and its stakeholders.

1. The rational agreement between who runs the firm and the non-controlling stakeholders which defines the abstract and general principle T and the related standard of CSR:
   - allow the firm to develop a reputation that can induce stakeholders to enter in cooperative relations with it;
   - make possible the creation of the ideal utility (the principle T has a key role in the utility function of agents with conformist preferences);
   - allow the formation of stakeholders' belief in the conformity of the firm to the principle T.
2. Beliefs and dispositions to conform to the CSR ideal principle T affect the ideal utility which depends on λ (when λ_i increases, it increases also the utility of the stakeholder i in cooperating with a firm that conforms to the principle T) and on the belief on the others' behaviour with respect to the conformity with the principle T.
3. The ideal utility reduces the cost of developing a reputation. Stakeholders who obtain ideal utility by cooperating with a firm

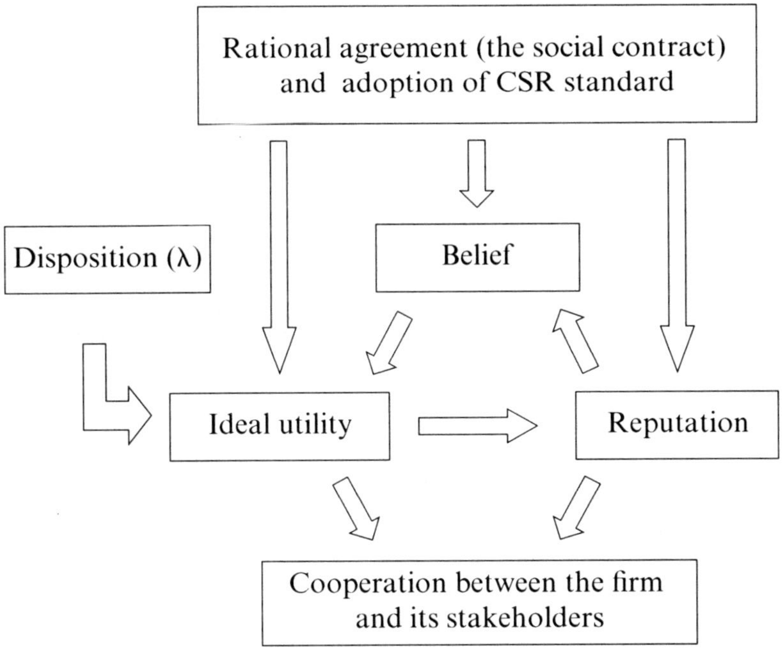

Figure 7.1 The relationship between cognitive social capital, adoption of CSR standards, reputation and creation of cooperative relations

which respects the CSR principle *T* will trust the firm sooner than stakeholders who do not obtain ideal utility.

4. A good reputation contributes to increase belief that the firm will respect the CSR principle. Stakeholders with conformist preferences who notice that the firm has a good reputation (because it respects its commitments) will reinforce their belief in the cooperative character of the firm.
5. Ideal utility and reputation induce stakeholders to enter into a cooperative relation with the firm. In particular, the ideal utility obtained by strong stakeholders in cooperating with a firm which conforms to the principle *T* could induce the firm to respect all the stakeholders. In fact, if the firm stops cooperating with weak stakeholders, the strong stakeholders lose their ideal utility in cooperating with it. If λ is high enough, the strong stakeholders may decide to punish the firm which abuses the weak stakeholders by stopping their cooperation with it. This possibility is a credible threat for the firm because it is connected with endogenous incentives related to the effect of cognitive SC on

stakeholders' ideal utility. The fear of being punished by the strong stakeholders can induce the firm not to abuse the weak stakeholders, thus making cooperative relations sustainable also between the firm and its weak stakeholders and generating structural SC.

5. CONCLUSION

The aim of this chapter was to investigate the relationship between social capital and CSR. Our principal point has been to highlight the importance of cognitive SC, understood as dispositions and beliefs, and CSR principles in generating networks of cooperative relations between the firm and its stakeholders. In order to obtain this goal, we have introduced a definition of CSR coherent with a contractarian approach and two different notions of SC, the cognitive and the structural social capital. Cognitive SC has been defined as dispositions to conform with ethical principles of cooperation and as beliefs in other conformity to the same principles. With regard to the CSR notion, it has been defined as a model of extended corporate governance whereby those who run a firm have responsibilities that range from fulfilment of their fiduciary duties towards the owners to fulfilment of analogous fiduciary duties towards all the firm's stakeholders. We have also introduced a distinction between strong and weak stakeholders. The former bring strategic assets into the firm, the latter invest in the firm but without bringing in strategic assets. The firm is interested in cooperating in the long term with strong stakeholders and it is not interested in doing the same with weak stakeholders.

According to our theoretical argument, the disposition of stakeholders to conform with agents who share ethical principles of cooperation represents a incentive for the firms to adopt CSR standards. Stakeholders who are characterized by this kind of disposition will trust a firm which respects ethical principles of cooperation sooner than stakeholders who are not. For this reason, to declare (and to observe) CSR principles is more convenient for a firm who acts in a context characterized by a high level of cognitive social capital (in terms of dispositions) than for a firm who is connected with stakeholders who are not endowed by cognitive social capital.

In order to arouse trust in stakeholders, because of the incompleteness of contracts and bounded rationality, the firm has to explicitly declare the principles (in terms of fulfilling of the fiduciary duties towards all the stakeholders) that will characterize its behaviour. By comparing the statements, formulated in the CSR standard, and its behaviour, the stakeholders can check if the firm actually respects its duties. The CSR standard is essential in order to allow stakeholders to form their beliefs in the conformity of the firm to the ethical principles of cooperation and to decide if they may trust the firm.

After dispositions have fostered the diffusion of CSR principles and the CSR standard has activated the beliefs, we will observe that stakeholders (both weak and strong stakeholders) start a cooperative relationship with the firm and start to undertake investment at an optimal level in the firm.

According to our idea of conformist preferences, stakeholders characterized by cognitive SC do not get only an economic payoff by cooperating with a firm who conforms with ethical principles of cooperation. They also get an ideal utility that originates from the fact of cooperating with a firm who respects the principles with which they want to conform.

If the firm behaves opportunistically with one or more stakeholders the ideal utility goes to zero for all the stakeholders, because they observe that the firm stops conforming to the ethical principles of cooperation. If the loss of the ideal utility is high enough, the stakeholders may decide to interrupt their cooperation with the firm (and maybe to start a new relation with another firm which does respect the principles). This possibility does not worry the firm when it concerns only the weak stakeholders. On the contrary, the interruption of the cooperation may generate a economic loss for the firm if it is decided by strong stakeholders.

On the basis of these premises, the reasoning developed in section 4.3 has shown that the cognitive SC of strong stakeholders, associated with the adoption of CSR practices by the firm, may avoid opportunistic behaviour of the firm against weak stakeholders, even though, in each single relationship with weak stakeholders, the firm would have economic incentive to defect. Essentially, the possibility that strong stakeholders decide not to cooperate with the firm if it defects with weak stakeholders is a reliable threat for the firm that may decide (it depends on the payoff structure) to cooperate with weak stakeholders in order to avoid sanctions from strong stakeholders.

In conclusion, our analysis identifies the conditions for a virtuous circle between cognitive SC, CSR and structural SC. In particular, dispositions, beliefs and CSR practices make sustainable cooperative relations between the firm and its weak stakeholders that would not be sustainable otherwise. This is a socially desirable result because: (1) all the stakeholders undertake investment at an optimal level in the firm; (2) the cooperation between the firm and all its stakeholders is successful and generates a higher total output than the total output that would be generated by opportunistic behaviour.

APPENDIX 1

In this appendix we focus on the function F which is a function, shared by all the agents, of the social normative criterion T.

Following Grimalda and Sacconi (2002), we adopt a particular specification for F based on an idea of expected mutuality in conforming to the normative prescriptions. Grimalda and Sacconi (2002) restrict the attention to a two-person game and define two indices that contribute to determine F:

1. f_i: the index of conditional conformity of player i (or degree of deviation from pure conditional conformity with T):

$$f_i(\sigma_i, b_i^1) = \frac{T(\sigma_i, b_i^1) - T^{MAX}(b_i^1)}{T^{MAX}(b_i^1) - T^{MIN}(b_i^1)}$$

   Where $T^{MAX}(b_i^1)$ and $T^{MIN}(b_i^1)$ are respectively the maximum and minimum value that the welfare distribution function, representing the normative principle or ideology, can assume, depending on i's action, given i's first order belief, b_i^1, over the action that j is going to perform.
   $T(\sigma_i, b_i^1)$ is the actual level of T when player i implements strategy σ_i given what s/he expects from player j.
   f_i varies from 0 (no deviation at all from the principle T) to -1 (maximal deviation).
2. $\bar{f}_j$ is the esteem that player i forms about j's compliance with the ideology.

$$\bar{f}_j(b_i^1, b_i^2) = \frac{T(b_i^1, b_i^2) - T^{MAX}(b_i^2)}{T^{MAX}(b_i^2) - T^{MIN}(b_i^2)}$$

   b_i^1 is the first order belief that player 1 has in the action of player j. b_i^2 is the second order belief about player j's belief in the action adopted by player i. $T^{MAX}(b_i^1)$ and $T^{MIN}(b_i^2)$ are the value that the welfare function takes when player j respectively maximises or minimises it, given the second order belief of player i. In other words, those functions indicate the maximum and minimum value that player j can attribute to the welfare function, given the belief s/he has about i's action as perceived by i himself/herself. $T(b_i^1, b_i^2)$ is the actual value that i expects the welfare function to take according to his beliefs. $\bar{f}_j$ varies between 0 and -1 that respectively indicates the maximum and minimum degree of conformity by player j to the ideology as embodied in the welfare function T.

Implementing these definitions, the utility function of agents can be written:

$$V_i(\sigma_i, b_i^1, b_i^2) = U_i(\sigma_i, b_i^1) + \lambda_i[1 + \bar{f}_j(b_i^1, b_i^2)][1 + f_i(\sigma_i, b_i^1)]$$

ACKNOWLEDGEMENT

We would like to thank Provincia Autonoma di Trento for financial support in the project 'Social capital, corporate social responsibility and local economic development'.

NOTES

1. It is possible to identify two principal approaches to SC: social capital in terms of generalised trust or civic norms (e.g. Putnam et al., 1993 and Knack and Keefer, 1997) and in terms of social networks (e.g. Coleman, 1988; Lin, 2001; Burt, 2002).
2. These relational elements related to the networks among firms and stakeholders are indubitably less important with regard to the implementation of CSR practices if one considers other approaches to CSR. This is the case, for example, of the approaches by Friedman (1977) and Jensen (2001). Both these authors do not give much room to the explicit consideration of the stakeholder's interests by the owners of firms (see section 3).
3. On the concept of fiduciary duty see also Flannigan (1989).
4. Specific investments may significantly increase the total value generated by the firm (net of the costs sustained for that purpose) and are made in relation to a specific firm (and not in any other).
5. The decision about the party that must have the residual right of control may depend on various factors – e.g. a comparative analysis of control's costs of the different stakeholders – see Sacconi, 2006 for a deeper explanation.
6. There are two main fiduciary duties towards the non-owners that should be defined in the social contract:
   1. To avoid the production of negative external effects on stakeholders not party to transactions, or compensate them so that they remain neutral;
   2. To remunerate the stakeholders participating in the firm's transactions with payoffs which, taken for granted a fair status quo, must contain a part tied to the firm's economic performance such to approximate fair/efficient shares of the surplus (assuming that this is positive) as envisaged by the first social contract.
7. With regards to the rational bargaining over the firm constitutions and the related Nash bargaining solution (Nash, 1950; Harsanyi, 1977) see Sacconi, 2006.
8. This theory of reputation under unforeseen contingencies is fully developed in Sacconi (2000 and 2004). For a design of a CSR management standard that corresponds to the characters now defined see, for example, Sacconi et al. (2003) and Clarkson Centre for Business Ethics (2002).
9. See Appendix 1 for a formal representation of F.
10. Moreover, in a previous work, Sacconi (2006) has shown that if stakeholders have conformist preferences, the firm cannot apply a mixed strategy or a 'refined abuse strategy'

(i.e. a strategy which induces an equilibrium in which the firm abuses with the maximum possible probability compatible with maintaining stakeholder indifference between entry and non-entry).

REFERENCES

Baron, D.P. (2005), 'Corporate social responsibility and social entrepreneurship', Stanford, Graduate School of Business research paper series no. 1916.

Burt, R. (1992), *Structural Holes*, Cambridge, MA: Harvard University Press.

Burt, R. (2002), 'The social capital of structural holes', in M.F. Guillen, R. Collins, P. England and M. Meyer (eds), *The New Economic Sociology*, New York: Russell Sage Foundation.

Clarkson Centre for Business Ethics (2002), 'Principles of stakeholder management', *Business Ethics Quarterly*, **12** (2), 257–64.

Coleman, J.S. (1988), 'Social capital in the creation of human capital', *American Journal of Sociology*, **94**, 95–120.

Coleman, J.S. (1990), *Foundations of Social Theory*, Cambridge, MA: Harvard University Press.

Flannigan, R. (1989), 'The fiduciary obligation', *Oxford Journal of Legal Studies*, **9**, 285–94.

Freeman, E. (1984), *Strategic Management, a Stakeholder Approach*, Boston, MA: Pitman.

Freeman, E. (2000), 'Business ethics at the millennium', *Business Ethics Quarterly*, **10** (1), 169–80.

Freeman, T. and W.M. Evan (1990), 'Corporate governance: a stakeholder interpretation', *Journal of Behavioural Economics*, **19** (4), 337–59.

Freeman, R.E. and J. McVea (2002), 'A stakeholder approach to strategic management', Darden Graduate School of Business Administration working paper no. 01-02.

Friedman M. (1977), 'The social responsibility of business is to make profits', in G.A. Steiner and J.F. Steiner (eds), *Issues in Business and Society*, New York: Random House.

Grimalda, G. and L. Sacconi (2002), 'The constitution of the nonprofit enterprise: ideals, conformism and reciprocity', *Liuc Papers n. 115, Law and Economics Series*.

Grimalda, G. and L. Sacconi (2005), 'The constitution of the not-for-profit organization: reciprocal conformity', *Constitutional Political Economy*, **16** (3), 249–76.

Grossman, S. and O. Hart (1986), 'The costs and benefit of ownership: a theory of vertical and lateral integration', *Journal of Political Economy*, **94**, 691–719.

Hansmann, H. (1996), *The Ownership of Enterprise*, Cambridge, MA: Harvard University Press.

Harsanyi, J.C. (1977), *Rational Behaviour and Bargaining Equilibrium in Games and Social Situations*, Cambridge: Cambridge University Press.

Hart, O. (1995), *Firms, Contract and Financial Structure*, Oxford: Clarendon Press.

Hart, O. and J. Moore (1990), 'Property rights and the nature of the firm', *Journal of Political Economy*, **98**, 1119–58.

Jensen, M.C. (2001), 'Value maximization, stakeholder theory, and the corporate objective function', *Journal of Applied Corporate Finance*, **14** (3), 8–21.
Knack, S. and P. Keefer (1997), 'Does social capital have an economic payoff? A cross country investigation', *Quarterly Journal of Economics*, **CXII**, 1251–87.
Lin, N. (2001), *Social Capital*, Cambridge: Cambridge University Press.
Nash, J. (1950), 'The bargaining problem', *Econometrica*, **18**, 155–62.
Paldam, M. (2000), 'Social capital: one or many? Definition and measurement', *Journal of Economic Surveys*, **14** (5), 629–53.
Putnam, R., R. Leonardi and R. Nanetti (1993), *Making Democracy Work: Civic Traditions in Modern Italy*, Princeton, NJ: Princeton University Press.
Sacconi, L. (1999), 'Codes of ethics as contractarian constraint on abuse of authority: a perspective from the theory of the firm', *Journal of Business Ethics*, **21**, 189–202.
Sacconi, L. (2000), *The Social Contract of the Firm, Economics, Ethics and Organisations*, Berlin: Springer Verlag.
Sacconi, L. (2004), 'Incomplete contracts and corporate ethics: a game theoretical model under fuzzy information', in F. Cafaggi, A. Nicita and U. Pagano (eds), *Legal Orderings and Economic Institutions*, London: Routledge.
Sacconi, L. (2006), 'A social contract account for CSR as an extended model of corporate governance (I): rational bargaining and justification', *Journal of Business Ethics*, special issue on social contract theories in business ethics, 259–81.
Sacconi, L. (2007a), 'A social contract account for CSR as an extended model of corporate governance (II): compliance, reputation and reciprocity', *Journal of Business Ethics*, **75** (1), 77–96.
Sacconi, L. (2007b), 'CSR as a model of extended corporate governance, an explanation based on the economic theories of social contract, reputation and reciprocal conformism', in F. Cafaggi (ed.), *Profiles of Self-regulation*, Kluwer Academic Press.
Sacconi, L., S. DeColle and E. Baldin (2003), 'The Q-RES project: the quality of social and ethical responsibility of corporations', in J. Wieland (ed.), *Standards and Audits for Ethics Management Systems, The European Perspective*, Berlin: Springer Verlag, pp. 60–117.
Uphoff, N. (1999), 'Understanding social capital: learning from the analysis and experience of participation', in P. Dasgupta and I. Serageldin (eds), *Social Capital: A Multifaceted Perspective*, Washington, DC: World Bank.
Williamson O. (1975), *Market and Hierarchies*, New York: Free Press.
Williamson O. (1986), *The Economic Institution of Capitalism*, New York: Free Press.

8. Creativity and institution building: the case of Italian social cooperatives

Alberto Ianes and Ermanno Tortia

1. INTRODUCTION

The dismal science of economics and its tools – institutions, the market, contracts, and the enterprise in its multiple forms, can benefit from creative capacity when cooperative firms are considered. More than a century of history has seen creativity give rise to new cooperative institutions in Italy furnishing original solutions to emerging needs. In this way, the cooperative enterprise has been ahead of its times and anticipated both the market and the law, which has often ratified *ex post* a phenomenon already fully-fledged.

One of the most recent examples of this creativity has been the social cooperative, a type of organization developed in the 1970s to innovate social policy in Italy. It shifted the debate to hypotheses hitherto considered fanciful – or one might say 'creative'. The idea was to combine in the same organization, solidarist aims (i.e. concerned with the collective interest) and entrepreneurial activity, with business risk – besides managerial power – being voluntarily and directly assumed by workers, volunteers, and the users of services.

In this regard, whilst 'economic theory' has only just begun to reflect on forms of enterprise not managed in the exclusive interest of the ownership, business history has entirely failed to address an entrepreneurial model with these distinctive characteristics. Yet we believe that only the joint use of different but contiguous disciplines,[1] history and economic theory, can yield interpretation of a phenomenon, that of the social enterprise, whose sustainability and reproducibility is difficult to construe using standard tools.

It is our conviction that the joint use of history and economic theory can only be fruitful, and therefore only able to furnish a convincing explanation of the existence and workings of the social enterprise, if it is accompanied by the willingness of both disciplines to agree on certain principles. It is particularly important that history should cease being merely the description of dates, facts and events and also put forward interpretations. Economic theory, for its part, must furnish interpretative tools which bring historical events as close as possible to reality. Hence an explanatory paradigm must assume more concretely-grounded hypotheses. It is precisely this approach that is adopted by this chapter.

The second section describes the history of Italian social cooperatives interpreted as the outcome of a process of institution building driven by various factors: a change of context; entrepreneurial and social innovation brought about by volunteers; and the creative imagination of social cooperators endowed with organizational and governance architectures consistent with the characteristics of the enterprise and the needs of the sector.

The third section interprets the social enterprise using the tools furnished by evolutionary theory, which today, probably, is the theory best able to grasp the context and its evolution – its history one might say – and the different types of enterprise. It is through a process of innovation, imitation and selection that, according to the evolutionist theory, new business institutions and more consistent organizational routines are created. The evolutionary interpretation enables Italian social cooperation to be framed within a broader theoretical and historical context. First, social cooperatives can be considered an instance of the wider family of social enterprises recently introduced in national legislations[2] and which are spreading in countries such as the United States. Second, the distinctive patterns of interaction that characterize organizational routines in social cooperatives allow the study of specific evolutionary mechanisms – for example concerning motivations and organizational objectives – that can increase knowledge about the involvement of workers and other actors (users, volunteers, and the local community).

2. INSTITUTION BUILDING AND SOCIAL ENTERPRISES FROM A HISTORICAL PERSPECTIVE

Business history, even in its streams most sensitive to the Chandlerian thesis of the diversified and multidivisional firm (Chandler, 1977, 1990), has completely disregarded enterprises characterized by goals other that the maximization of profit. Nor has it been envisaged that the owners of a business organization may differ from stockholders in that they pursue other interests, including solidarity and altruism.

And yet there exists at least one form of enterprise – namely cooperation in its traditional forms – which pursues goals other than the maximization of profit and, moreover, whose owners are not stockholders. The owners of worker cooperatives are in fact the workers, those of consumer cooperatives are the consumers, and those of agricultural cooperatives are the pools of farmers. However, since the Civil Code compiled in 1942, Italian cooperatives have been exclusively mutualistic, i.e. constituted to improve the conditions of their members, not of the community at large.

A change intended to reconcile the interests of business and the local community came about in the 1970s, when there arose a new cooperative form – the social cooperative – whose principal aim was to pursue the community's general well-being by supplying social welfare services and employing disadvantaged persons (Borzaga and Ianes, 2006; Borzaga and Santuari, 2001). In this regard, the creation of social cooperatives innovated the very conceptions of the 'nature of the firm'.

The social cooperative has revolutionized the conception of the for-profit enterprise because it comprises two apparently incompatible aspects – business and solidarity – in pursuit not of profit but the general interest of the community. It also differs from traditional companies because, although it is a business, it performs a typically statist function, that of the voluntary distribution of wealth according to a principle other than equivalence. This function is performed not through the transfer of resources, but directly through the supply of goods and social services to persons without sufficient economic means to purchase them. This is possible because involved in social enterprises are actors, such donors and volunteers, who perform an important though non-remunerated role. The workers themselves receive a wage less than that warranted by the effort put into the organization. Yet, as some empirical studies have shown (Benz, 2005; Borzaga and Depedri, 2005; Borzaga and Tortia, 2006), they declare themselves satisfied with their work mainly because they are driven by intrinsic motivations: the possibility to participate in decisions as owners, to govern the organization, or to feel themselves actively involved in a project which they perceive as their own and which they value. This enables the social cooperative to obtain skilled human resources which behaves in a manner coherent with the organization's strategies because these are endorsed by the workforce and by the management.

The history of the Italian social cooperative shows how it is possible to adapt the Schumpeterian notion of the innovating entrepreneur to the cooperative and thus yield understanding of the creativeness of 'innovating cooperators'. The cooperative has adjusted itself to new demands, and from mutualistic it has made itself solidarist.

2.1 Attention to the Context

The social cooperative did not come into being by chance; rather, it did so in close causal relation to a context – that of the 1970s – undergoing radical change. In that decade, in fact, Italian society and economy were subject to strong tensions due to transition from an industrial society to one difficult to define. This transition had three main features: unemployment, tertiarization, and the crisis of the welfare state (Borzaga and Ianes, 2006).

First, unemployment, consequent on the economic crisis, hit two extreme components of the labour market: young job-seekers, mainly female, and workers close to retirement age who risked losing their jobs before they had made the contributions necessary to qualify for a pension. Unemployment was the consequence of the 'hot autumn' of 1969 and of the two oil shocks of 1973 and 1979. In the autumn of 1969, in fact, action by the trade unions brought gains for their rank and file, such as wage increases and improved working conditions. But it also increased the cost of labour and pushed up prices. Likewise, the increased cost of raw materials, particularly of oil, intensified inflationary pressures, which persisted throughout the 1970s and thereafter. To curb price rises, the monetary authorities and the government adopted restrictive monetary and economic policies. But these measures did little to bring inflation back to normal levels. Instead, they triggered stagflation, a fall in demand, and a general slowdown of the economy, and – as said – a worrying increase in unemployment.

Another phenomenon indicative of the change ongoing in the 1970s was tertiarization, which altered the composition of demand. Demand for consumer durables progressively diminished, with a concomitant increase in that for personal and family services. Various factors were responsible for this phenomenon. First, the Italian family changed from being patriarchal to nuclear, with a consequent reduction in the number of its members. Concommitantly, female employment started to increase. This new kind of family found it difficult to continue furnishing 'home care' for its children and elderly members as it had previously done.

Another cause of the increase in demand for personal services was the advent of the so-called 'new poverty' or 'post-materialistic poverty': this was produced by spread of drug and alcohol abuse, and other problems related to the difficult circumstances of Italy's urban hinterlands.

New needs also arose from the achievement of certain social advances. For example, law 180/78 (the so-called 'Basaglia Law'), ordered the closure of all mental asylums in the country, implying that the problem of mental illness had been transferred to society as a whole. The network of social protection was unaccustomed to dealing with situations of hardship which were due, not to material circumstances, but to a person's mental state.

The situation was made even more problematic by a third factor: the above-mentioned crisis of the welfare state, whose weaknesses derived from a fiscal and financial crisis, and an ungoverned increase in the public debt due in large part to 'cavalier' management of government accounts for – at times – electoral purposes. But there was an even more serious crisis. This was of an organizational nature and stemmed from the rigidity of the public administration and its inability to respond efficiently and effectively to the new problems. These latter required closer attention to users, flexibility of action with service recipients, and a capacity to deal with the new needs of citizens. The state therefore proved inadequate in coping with 'new poverty' in Italy.

2.2 The Creative Process and the Social Enterprise

Little action was taken in response to the difficulty of the state and for-profit enterprises, but one initiative arose from that part of society most sensitive to new poverties, having acquired critical awareness from first-person experience of the upheavals of 1968, or of the Vatican Council II, which had contributed to renewing the Church. During the 1970s, social activists within this cultural context set up a number of voluntary schemes that were later demonstrated to be the initial step toward more structured and emancipated organizational solutions. Indeed, the crisis of the welfare state was not temporary but definitive, and it could not be resolved with the usual tools of the public welfare system.

These early volunteers consequently examined Italian law for a legal arrangement which could give greater stability and continuity to a production of services that had been provisional since the first schemes of organized voluntary work. In other words, they sought to identify a legal form able to reconcile two apparently incompatible aspects: solidarity and entrepreneurship. However, few solutions were forthcoming: according to the Italian civil code, associations and foundations, for example, could not undertake business activities.

The only legal form able in some way to reconcile being businesslike (that is, an efficient organization) with fiduciary relationships in pursuit of the community's general interest was the cooperative enterprise. Consequently, the early 1970s saw the first instances of social cooperation, those varyingly called social solidarity cooperatives, social services cooperatives, or integrated cooperatives. When legal recognition was granted in 1991, all these variants took the name of 'social cooperatives'.

The social cooperative can accordingly be interpreted as a process of learning by doing: that is, as the outcome of a creative process, of adaptation to the new needs of undertakings (cooperatives) already active in diverse areas of the economy: credit, consumption, the processing and storage of agricultural produce. In face of Italy's new needs, however, the cooperative changed its form from mutualistic to solidaristic.

The process was neither simple nor painless. The first social service cooperatives, in fact, often had ratification of their statutes denied by the courts on the ground that they breached the principle of mutuality, so that the cooperatives could not become operational.

Two devices were used by social entrepreneurs to gain recognition as cooperatives. The first consisted in appeal to article 45 of the Italian Constitution, which recognizes the 'social function' – and not the solely mutualistic function – of the cooperative enterprise. The second stratagem exploited an expression – 'external mutuality' or 'enlarged mutuality' – coined to extend the concept of mutuality from the exclusive interest of the cooperative's members to the general interest of the community. Thanks to these two devices, the first social service cooperatives obtained provisional permission to operate from the courts, which, on re-examining their statutes revised their opinions and ratified the statutes.

Yet all this was a contrivance: these new cooperatives were neither mutualistic nor homogeneous. The benefits deriving from businesslike management accrued largely outside the membership. Moreover, many of these organizations involved people with diverse aspirations: the social structure became multi-stakeholder. The business risk was assumed, as said, by stakeholders with differing aspirations and goals.

Awareness of the different nature of this new cooperative with respect to the traditional form stimulated a debate which lasted for fully ten years, and on conclusion of which the approval of law 381/91 gave social cooperatives a specific status and a recognized space of action. The social cooperative was recognized as an organization delivering collective and commonal services with the purpose of pursuing the general interest of the community.

In particular, article 1 of the law envisaged two possible types of social cooperation: 'type A' cooperation concerned with the 'management of social, health and educational services', and 'type 'B' cooperation concerned with integrating disadvantaged persons into work through the management of various activities: agricultural, commercial and industrial, as well as others like the management of services.

More recently, in 2005, the definition of social cooperatives has been included into a more general legal form termed the 'social enterprise'. Social enterprises retain the main features of social cooperatives, i.e. the social and public benefit objective, multi-stakeholder governance, and the profit distribution constraint, but the law expands the range of organizational options since they are not forced to assume mutualistic form (i.e. be cooperatives), but also traditional non-profits (i.e. associations and foundations) can be included under the definition together with investor-owned firms which fulfil the above requirements.[3]

2.3 Quantitative Elements

The evolution just described has been reflected in the quantitative aspects of the growth of social cooperatives. The first survey on the phenomenon was conducted in 1987, and it already highlighted the characteristic features of these organizations, but also their potential for growth and development. As of 31 December 1986, the 496 cooperatives surveyed had 4265 voluntary members. There were also 4057 worker members, 2277 non-member volunteers, and 704 employees. To these were added 2412 disabled members, and the 22684 users of the services supplied by the cooperatives.

At the time of approval of law 381/91, which regulated social cooperatives, their number was estimated at just over 100. The figure rose to around 3000 in the mid-1990s, and then grew further to 7363 as of 31 December 2005, as reported by the most recent national institute of statistics (ISTAT) survey (see Tables 8.1 and 8.2).

Over 70 per cent of the more than 7000 cooperatives surveyed in 2005 had been constituted since 1991, a figure indicative of the recent development of the phenomenon and its exponential growth – as noted by ISTAT – especially in the period 1996–2000. Out of the total of 7363 cooperatives, 59 per cent, equal to 4354 units, belonged to the type A category, in that they delivered social, health and educational services, whilst 32.8 per cent, corresponding to 2419 cooperatives, were of type B, and were therefore engaged in the work integration of disadvantaged persons, mainly physically, mentally and sensorially disabled. There were then 315 cooperatives which were mixed, in that they undertook the activities foreseen for type A and type B cooperatives, although, as will be seen, there were relatively few of them. Finally to be mentioned are 284 consortia, which represented 3.9 per cent of the total: these will be discussed in detail later in this chapter. In 2003 cooperative members numbered 262389, of which 255583 were individuals and 6806 legal entities, a 19 per cent increase on 2002. Another interesting finding with regard to the social base of cooperatives concerns the diverse types of members of which they were composed, and the consequent diverse forms of ownership. The survey results show that 81.1 per cent of cooperatives associated several categories of stakeholders, although only 21.1 per cent had more than three of them. This confirms

Table 8.1 Number of cooperatives and value of output by type in 2003 and 2005

	2003					2005				
	Type A	Type B	Category mixed (A+B)	Consortia of cooperatives	Total	Type A	Type B	Category mixed (A+B)	Consortia of cooperatives	Total
No. of cooperatives	3707	1979	249	224	6159	4345	2419	315	284	7363
Value of output (in millions of euros)	3107	1020	169	530	4826	4133	1354	215	680	6382

Source: our calculations on data from ISTAT (2006), *Le cooperative sociali in Italia, 2003*, information no. 30, Rome; ISTAT (2007), *Le cooperative sociali in Italia, 2005*, statistics in brief, Rome.

Table 8.2 Human resources, users and disadvantaged subjects of social cooperatives in 2003 and 2005

	2003	2005
Volunteers	31 879	34 626
Paid personnel	189 134	244 223
Users of Type A cooperatives	2 403 245	3 302 551
Disadvantaged personnel of Type B cooperatives	23 587	30 141

Source: Our calculations on data from ISTAT (2006), *Le cooperative sociali in Italia, 2003*, information no. 30, Rome; ISTAT (2007), *Le cooperative sociali in Italia, 2005*, statistics in brief, Rome.

a distinctive feature of the phenomenon: the most frequent ownership form is of multi-stakeholder type, even though the number of categories represented is small.

With reference to economic aspects, in 2005 the social cooperation movement achieved output amounting to around 6.4 billion euros, assuring paid employment for 244 223 thousand workers and involving more than 34 626 volunteers. The sector is largely female, given that 71.2 per cent of its personnel are women. Inspection of the main activities undertaken and the composition of users shows that in 2005 the majority (59.1 per cent) of type A cooperatives delivered social assistance, particularly in the home (36.5 per cent of cooperatives) and largely in the form of childcare, which accounted for 28.8 per cent of users. In 2005 the users served by Type A cooperatives numbered more than 3.3 million, a 37.4 per cent increase on the 2003 figure. By the same token, Type B cooperatives involved 30 141 disadvantaged persons in work entry schemes, recording a 27.8 per cent increase with respect to 2003.

This brief statistical survey testifies to the presence of a phenomenon which, though limited with respect to the economy as a whole, achieves levels of performance and growth, in terms of employment, services delivered, and users, often several times higher than those recorded by other business forms. This has enabled the social cooperative movement to put itself forward as an actor able to furnish satisfactory responses within Italian social policies.

2.4 Organizational Creativity and Enterprise Networks

A further key feature of the creative process involving social cooperatives is their organizational evolution, which has enabled them to assume the form of a system of enterprises (Ianes, 2008). In this regard, there is no universal institutional structure for social cooperatives, but instead different gradations in conceiving, devising and organizing them. For example, there may be cooperatives of small size, and others of large size with more than 40 employees.

Rather than expand, numerous cooperatives have preferred to spin off new initiatives; others have instead grown by undertaking new functions and activities. Among all the possible variants, the most original is indubitably the one based on the idea of the firms' network since social cooperatives place more emphasis on the collaboration among enterprises than on competition. The most mature form of integration among social cooperatives has been called the 'strawberry patch strategy' (Scalvini, 1989). At the basis of this approach is the notion that 'the development of strawberry plants is rapid, but no plant grows to more than a certain size. Each of them extends runners which take root at a certain distance to produce a new plant, which once it has reached a certain size, reproduces itself in the same way. Thus the entire patch is gradually covered by innumerable interconnected but self-sufficient plants' (ibid.). The metaphor of the strawberry patch can also be represented with the paradox of 'little-big size' whereby a system has several levels of integration, with multiple relations between entrepreneurial actors and society.

The first level consists of cooperatives interfacing with users and producing services that cater for specific social needs. These first-level organizations have three main features: specialization, small size, and close links with the local community. Specialization is regulated by an autonomously assumed code of behaviour: every social enterprise tends to focus on only one activity. If a need different from the one addressed arises, the cooperative does not expand its range of activity but promotes the birth of a new cooperative through a spinning-off process. The aim is to propagate new cooperatives from a parent cooperative, which supervises the spin-off both in the incubation phase and the more problematic one of start-up by furnishing resources and skills. Once the new cooperative has become fully fledged, it severs all formal connections with the cooperative from which it has been germinated, although it maintains collaborative relations with it.

The second feature is the strong bond with the local community. This is important to monitor needs and their evolution, to have accurate information about territorial dynamics, and to achieve strong local embeddedness, which is the sole source of trust and social capital. A third requirement of these organizations is that they should be of small size. This is essential for a climate of trust and collaboration to arise within the cooperative, which is hard to achieve in an organization of large size (more than 40 employees), given the difficulty of establishing meaningful relationships in large numbers of people. Obviously, small size on its own does not guarantee a climate favourable to trust: other elements are necessary as well, for instance, a predisposition to participation by the various stakeholders involved, and the adoption of governance mechanisms coherent with the goals pursued.

Although small size has advantages, it also imposes constraints which restrict a cooperative's ability to satisfy needs. Consequently, many social entrepreneurs have not forgone the opportunities offered by large size, in particular economies of scale. It is here that importance is assumed by a second level of integration consisting of second-degree consortia, the first of which was created in 1983. Consortia are largely provincial in their extent, and their purpose is to unite in a network system the majority of social cooperatives operating in a particular geographical area, regardless of their type of activity. In 1993 there were 31 consortial unions in Italy (Borzaga and Ianes, 2006). In 2001, as shown by Table 8.3, ISTAT counted 197 consortia, which increased to 284 in 2005, the year of the most recent statistical survey. Of these, 158 (equal to 55.6 per cent) of the latter operated in Northern Italy, 65 (22.9 per cent) in the Centre, while 61 were located in the South. In 2005 more than half of all social cooperatives belonged to a consortium. Moreover, each consortium comprised an average of 14 cooperatives (ISTAT, 2006, 2007).

Standing above the second-level consortia is the nation-wide Consorzio Gino Mattarelli (CGM) created in 1987 to give the system of social cooperatives an overall entrepreneurial and strategic vision in order to ensure the universality of services wherever social entrepreneurship operates.

Since its creation, the CGM has sought to aggregate an ever larger number of provincial consortia. In 2005, 79 local consortia belonged to the CGM, as opposed to 77 in 2004 and 69 in 2001, whilst the cooperatives belonging to the network via these consortia increased in number from 1033 in 2001 to 1156 in 2004 (Ianes, 2008). The organizational system headed by the CGM is illustrated in Figure 8.1.

According to this organizational scheme, the small size of the first-level cooperatives is a deliberate choice, because it favours profitable interaction with the local community, allows personalized responses, and facilitates effectively the democratic management of the firm. Likewise, the large size achieved through membership of a consortium is an equally deliberate choice, because it confers dynamism on the member cooperatives, broadens horizons, enables business growth processes to be managed, and gives cooperatives a certain bargaining power in their transactions with buyers and suppliers. Not all cooperatives belong to a consortia, just as not all consortia belong to the system of enterprises headed by the CGM.

Table 8.3 Number and size of consortia according to member cooperatives (and other units) by geographic area in 2001, 2003 and 2005

	2001			2003			2005		
	No. of members	No. of consortia	Members/ consortia	No. of members	No. of consortia	Members/ consortia	No. of members	No. of consortia	Members/ consortia
North	1822	120	15.18	1790	122	14.67	2504	158	15.85
Centre	381	39	9.77	526	56	9.39	751	65	11.55
South	482	38	12.68	783	46	17.02	820	61	13.44
Italy	2685	197	13.63	3099	224	13.83	4075	284	14.35

Source: Our calculations on data from ISTAT (2006), *Le cooperative sociali in Italia, 2001*, topic no. 30, Rome; ISTAT (2006), *Le cooperative sociali in Italia, 2003*, information no. 30, Rome; ISTAT (2007), *Le cooperative sociali in Italia, 2005*, statistics in brief, Rome.

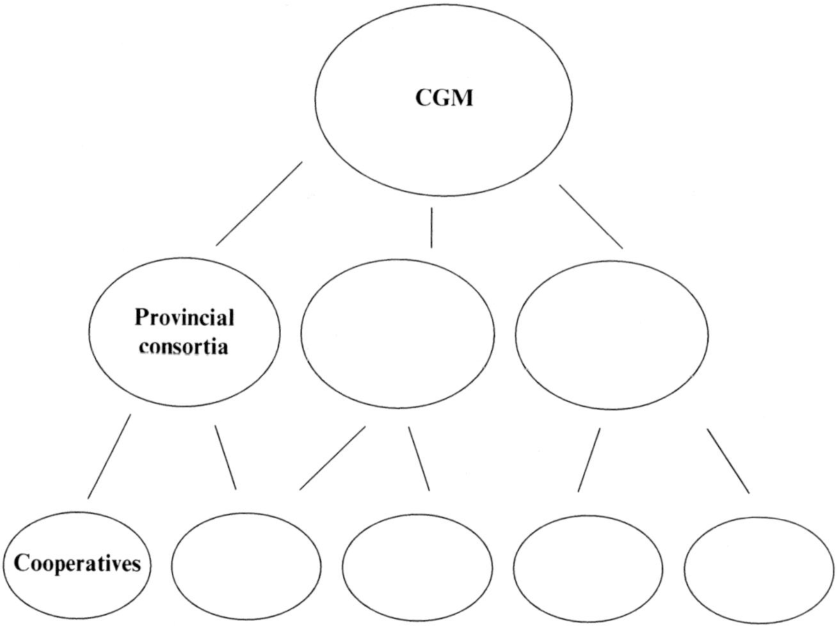

Figure 8.1 *Layered networking: structure of the social cooperatives' national and provincial consortia*

A good number of them, though, have joined a consortia. They have thus contributed their own expertise and profited from that made available by others, so that once again the reasons for cooperating prevail over those for competing.

2.5 The Efficiency of Social Cooperatives

As shown by empirical studies (Borzaga and Depedri, 2005; Borzaga and Tortia, 2006; Tortia 2007), social cooperation is organized in such a way as to produce trust, facilitate quality relationships among the actors involved, pursue both distributive and procedural equity, and activate a democratic decision-making system that motivates stakeholders. In regard to the economic-business side, whilst acknowledging that situations of difficulty exist, various studies have highlighted the ability of these organizations not only to balance their accounts but also to accumulate assets consisting of operating surpluses. Although the aim of social cooperatives is not to make a profit but to satisfy a need, the former is essential for undertaking new investments and creating forms of distribution able to support social services in a period when the state has major difficulties in coping with new and growing needs. These organizations are thus able to achieve better results in satisfying social needs than are for-profit firms, and they also gain a comparative advantage over the state because solidarity, democracy and equality become key factors in determining success. These are all factors which increase the endowment of social capital and an area's potential for development.

The features typical of social cooperatives, in fact, attract people to work for them less because they want to earn a wage and more because they share the organization's mission, feel involved, and are satisfied not by an income alone but by non-material rewards, too. This is reported by empirical surveys and the large body of theoretical literature on workers in non-profit organizations, and social cooperatives in particular (Borzaga and Musella, 2003; Borzaga and Depedri, 2005; Leete, 2000; Benz, 2005). Indeed, these studies maintain that lower average pay levels for workers in social cooperatives compared with those of public-sector employees correspond to higher levels of satisfaction for the former and lower ones for the latter. The same studies show that non-profit organizations and social cooperatives allow the 'selection' of persons – managers, employees and donors – who pursue efficiency by adopting behaviours coherent with the organization's objectives because they share them (Depedri, 2007).

The willingness of workers to accept wages lower than those of their counterparts in the public sector, the presence of volunteers in the organization, and the possibility of attracting economic and financial resources from donors, are all factors that boost production and help reduce costs without detriment to the quality of the service. This accounts for the existence of a form of enterprise which is difficult to explain by resorting to the canons of orthodox economic theory. Yet it is an enterprise whose history and quantitative aspects as described above have shown to be possible and real. Hence its existence requires a historical and social as well as economic explanation, and in this regard the evolutionary theory of economic change is likely to offer more innovative and flexible tools of analysis.

3. AN EVOLUTIONARY INTERPRETATION: VARIATION, DIFFUSION AND SELECTION OF SOCIAL COOPERATIVES

Contemporary evolutionary theory, since Nelson and Winter (1982), has developed important tools of analysis that can be fruitfully applied to the processes of creation and diffusion of new institutions and organizational forms, taking into account historical and developmental aspects usually disregarded by economic theory. This section is an initial attempt to apply the main concepts of evolutionary theory to the study of the emergence and diffusion of social cooperatives, which can be considered an instance of the more general category of social enterprises. This latter category is not restricted to Italy or Europe but represents an expanding entrepreneurial form in many advanced countries. Hence, the interest in this kind of evolutionary explanation.

3.1 Institutional Variation

As explained in the preceding sections, the origins of social cooperation can be traced back to social activism in Italy during the 1970s. The process of institutional creation underpinning the evolution of the new entrepreneurial form can be interpreted in Schumpeterian terms if creative destruction is not related exclusively to the introduction of innovative goods and services, but more broadly to the devising of innovative organizational routines and models. Indeed, the founders of social cooperatives – mainly in the initial years but also today – created both innovative services and organizational models which increased welfare and shifted the system from Pareto-inferior to Pareto-superior outcomes, as well as a wholly new governance system which, for the first time in Italy, integrated different actors in the governance of the organization. The process of creating and developing multi-stakeholder organizations, as ascertained by various in-depth case studies, was complex and required the creation of a network of actors involved step by step in the governance structure, so that, in the best examples, new social cooperatives arose as networks of co-motivated actors (Sacchetti and Tortia, 2008).

In its initial stages the new phenomenon was largely unnoticed by policy-makers and economic theorists, since these organizations were run mainly by volunteers, had a marginal role in the overall workings of welfare systems, and could not be explained on the basis of the traditional assumptions and analytical tools of economic theory. From an evolutionary perspective, it is clear that a process of variation (Hodgson, 2006), and hence of innovation at the institutional level, was taking place. This process was driven by both exogenous and endogenous factors. The endogenous factors consisted of the socially-laden set of values and ideologies inspiring the founders of the new organizations. Social sensibility, other-regarding presences (Ben-Ner and Putterman, 1999), and at times religious affiliation, were necessary preconditions for social activism and their absence would have precluded the starting up of the process. On the other hand, exogenous factors contributed as well, because the shortcomings of public supply, the crisis of the central welfare system, and the unwillingness of for-profit firms to enter the new sectors opened the way for social activists to launch their initial schemes. The new objectives to be pursued, far away for commercial ones, required new organizational tools and routines designed primarily to satisfy collective needs (Borzaga and Tortia, forthcoming). Institutional variation was justified by both subjective and objective conditions and concerned overtly organizational routines. This is why the development of social cooperatives offers a paradigmatic case for the analysis of economic and institutional evolution.

3.2 The Diffusion of New Institutional Solutions

The importance of this process of institutional variation is confirmed by its tendency to spread in the socio-economic system. The diffusion of new institutional contrivances is the second building block of an evolutionary theory of social cooperatives. It can be easily reconciled with similar processes concerning social enterprises then taking place in other European and North American countries. Both in Europe and in North America, non-profit organizations, social cooperatives and latterly social enterprises have grown in number and economic importance since the middle of the 20th century. Their weight in the economic system increased more rapidly than that of the other organizational forms operating more traditional activities. They grew mainly in the service sector, which is nonetheless the most dynamic part of advanced economies.

New organizational routines spread in a complex way. In general terms, diffusion takes place mainly by imitation. However, the modalities by which imitation is implemented require more in-depth analysis. Imitation is reported to be crucially influenced by a tacit dimension (Polany, 1958). Tacit knowledge and know-how cannot be learned solely through written or, more generally, codified information, not even in the presence of new and powerful information technologies. These can augment the quantity and quality of the information transmitted, making it more flexible and precise, but they cannot teach propensities crucial for the accomplishment of entrepreneurial tasks in unpredicted contingencies (Kreps, 1990). Organizational routines are similar to individual propensities, as they represent interlocking behavioural equilibria involving a plurality of subjects inside organizations (Hodgson, 2006). Hence, the transmission of these routines requires personal, often prolonged, interaction. Since organizational routines like individual habits are mere potentialities that do not need to become manifest, and because they are triggered by specific environmental stimuli which are at least partly unpredictable, codified knowledge can never convey all the information necessary to act appropriately in all possible events. Only learning by doing – and hence experience of a personal and relational nature – is able to create similar habits and routines suitable for dealing with similar situations.

Given these premises, it is not surprising that social cooperatives have developed and spread only in specific cultural contexts able to support the proper learning of habits and behavioural rules, i.e. institutional learning. This is a feature shared by many other social phenomena. The Industrial Revolution itself took place in the distinctive cultural context of eighteenth century England after the Glorious Revolution. In a similar fashion the building of trust relations and of fair organizational routines is particularly crucial for (though not at all unique to) the development of social enterprises. Whilst the replication of successful routines in the case of for-profit firms is often centred around the proper implementation of hierarchical control and economic incentives, in the case of social enterprises trust and fairness are more important, because hierarchy, control and monetary incentives are likely to be of limited effectiveness (Borzaga and Tortia, forthcoming). The difficulty of monitoring worker performance and the need for autonomy in carrying out tasks require a more inclusive and democratic governance where worker effort is driven primarily by intrinsic motivations, not by mere rewards and punishments. For these reasons, the learning process required to replicate organizational routines is likely to be particularly complex, and the tacit dimension paramount. Personal relations are crucial because it is first of all necessary to identify the main habits and propensities of the individuals involved, or at least of those with a strategic decision-making role. This first stage of learning, which primarily concerns individual behaviour and attitudes, is followed by a second stage where individual behaviours are combined within the framework of common objectives and procedures transmitted from individual to individual and from group to group inside the firm. Routines are more a feature of an organization than of a single individual or a group, and they can be transmitted from one generation of workers and managers to the next. Whilst hierarchy and control are made up of simpler and more easily imitated routines, trust and fairness are more difficult to imitate because they require voluntary engagement and satisfaction on the job, which are less crucial in the case of hierarchical organizational models. Trust in superiors together with procedural fairness are reported to be the main determinants of worker satisfaction (Tortia, 2008) in all ownership and organizational forms, and are indeed crucial in for-profit firms as well. However, they are a condition *sine qua non* for the firm's operation where non-hierarchical organizations are concerned. Hence the imitation and learning processes for social enterprises must start from the development of trust and from the implementation of fair procedures. The involvement of intrinsically motivated actors is no longer a possibility, but a necessity.

A cooperative attitude and its fit with organizational routines must be apprehended before any other aspect. This is more likely to happen when the organization has developed in value-laden social environments, as in the case of religious groups and social activism. The learning of, and conformity with, the set of values that inform the operation of the firm is often a precondition for the creation of an organization, and the evolution of the routines regulating the organization's behaviour must be coherent with its basic values and history.

As for the learning of the ability to take autonomous decisions in unpredicted contingencies, it usually involves a long process of trial and error which is not necessary when workers have to carry out standardized tasks on the basis of codified information and procedures. Individual ability to take autonomous decisions can of course have innate components as well. However, like any other propensity, it can be learned and implemented to varying degrees in the proper organizational context where participation is relevant. Procedural fairness and good relations with managers can be important in this regard, since they help improve workers' perception that they can act autonomously and accomplish tasks whose outcomes will be properly assessed by their superiors. Procedural fairness assumes the role of a structural feature of the organization, whilst interaction with superiors conveys the importance of personal relations and embeddedness.[4]

3.3 Selective Processes Facing the Diffusion of Social Enterprises

The last element in constructing an evolutionary theory of social cooperatives is the selective pressure applied by the surrounding socio-economic environment. Selection is crucial in any evolutionary explanation because resource scarcity and the availability of competing alternatives always apply pressure on existing and newborn organizations. Both similarities and differences can be envisaged in this realm with respect to for-profit firms, whilst radical differences are apparent with respect to public bodies.[5] Competitive pressure on markets or quasi-markets characterized by the sale of different types of services can be a feature shared by social cooperatives and for-profit firms. However, there are major differences in the way firms compete or cooperate, in their sources of finance, and in their allocation of resources. While the operation of for-profit firms is centred around the maximization of economic surpluses, and cooperative behaviour at the individual, organizational, and inter-organizational levels is instrumental to the pursuit of this aim, the reverse of this causal connection can be observed in the case of social enterprises (Borzaga and Tortia, 2007, forthcoming). Cooperation based on fairness and reciprocity becomes the norm, while competition and the pursuit of economic objectives are instrumental to the survival and expansion of the firm. The main economic objective is achieving equilibrium between costs (operating costs plus capital accumulation) and revenues, not maximization of the surplus. Competition can support a higher degree of efficiency and innovation, but it should be mainly interpreted as a process that expands variety and opportunities. Its aim is to introduce different organizational and productive solutions able to increase users' welfare, and not competition to increase profits and market shares.

There are remarkable differences in revenue sources as well. The revenues of for-profit firms derive mainly from sales of goods and services on the market, even if sales to public bodies are important as well, at least in some sectors. Social enterprises can instead draw on a wider set of sources, ranging from public subsidies to private donations, even if commercial revenues are growing in importance in most countries when non-profit organizations are considered (Weisbrod, 1998). Public support in the form of the purchase of goods and services, fiscal advantages and subsidies are integral parts of the revenues expected by social enterprises, more so than in the case of for-profit firms. This is due to a variety of reasons. First, the partially public and meritorious nature of the services produced include social enterprises among the most 'natural' suppliers for public bodies. Second, the social aim and the non-profit constraint reduce the risks of morally hazardous behaviour in regard to uninformed buyers or hard-to-monitor service quality. Because the bureaucratization of controls and risks of bribery often prevent public bodies from maintaining proper control over the quality of the goods and services produced, social enterprises are in a better position to create trust relations based on non-exploitative long-term contractual relations. Third, the ability to involve a variety of actors different from investors in the governance of the organization is a distinctive feature lacking in for-profit firms. Participation can be conducive to obtaining increased financial support in the form of stricter contractual relations with public bodies, and donative behaviour both in terms of monetary resources and labour (Borzaga and Tortia, 2007). Differences in revenue sources have important allocative and distributive consequences as well. Whilst for-profit firms are primarily devoted to increasing private gains, this is not the case of social cooperatives, whose social aim supported by the multi-stakeholder structure is intended to be conducive to a fair distribution of resources among all the actors involved. The exclusion of the profit motive triggers more complex distributive patterns whereby different distributive equilibria can be envisaged in different sectors of activity and in different organizations (Tortia, forthcoming). A unique or simple solution is not forthcoming, and different distributive patterns depend on a variety of factors, such as the assignment of control rights, the involvement of different actors, the fulfilment of the social aim, and the nature of the costs of producing specific good and services. Whilst a simple analytical derivation of distributive outcomes is excluded, more interesting questions can be asked about distributive patterns, for example about the institutional factors that regulate them and about the empirically observed outcomes.

The differences in financial possibilities should be added to the differences in the scope and importance of revenue sources. Social cooperatives are clearly at a disadvantage with respect to for-profit firms in that access to financial markets and to other sources of equity is barred because control rights are not assigned to investors, and credit by the banking systems may be restricted if proper collateral is not available. They rely mainly on internal funding in the form of self-finance through accumulated surpluses and on financial donations, together with financial support from the banking system.

The marked differences in revenue and financial sources support the idea that selective pressure operates differently in the cases of for-profit firms and social cooperatives. Whilst the former mainly undergo selection in terms of market competition for increased profits and market shares, the latter are subject to a more complex selection process. Market pressure is likely to be relatively weak in sectors where the public and meritorious nature of the services produced reduces effective demand and does not encourage entry by firms seeking high margins. On the other hand, reputation, service quality and the ability to certify the social impact of the firm's operation acquire paramount importance in the eyes of users, volunteers, donors, and the local community. Hence the quality of the services produced and their certified social impact are the main criteria in defining the firm's survival and expansion possibilities.

3.4 A Synthesis

To end this section, it is important to piece together the three components of the evolutionary process and explain how they interact. Innovation in terms of institutional variation conducive to the introduction of new organizational routines may take place at the individual or group level, but the cultural context and the presence of value-laden ideologies seem crucial in informing the development of solutions able to support the pursuit of relevant social aims. Organizational routines not only perform the function of underpinning innovative propensities and behaviour; they also replicate this behaviour and transmit it in a largely tacit way. They determine the success possibilities of the new solutions by way of interaction with the external socio-economic environment, which exerts selective pressure on the firm. Innovation is carried over to other organizations and groups of individuals through an imitative process that requires personal interaction, the transmission of tacit knowledge and the recognition of common values and objectives (Sacchetti and Sugden, forthcoming). The processes of diffusion and selection should be kept separate. Whilst the former process operates from within the organization and requires the learning and sharing of common values, the latter operates from outside and serves to test the quality of final outcomes. Of course, the three stages of the evolutionary process interact, because the quality of new institutional solutions and of diffusion processes impact on survival possibilities in the face of selective pressure. However, survival possibilities are never defined univocally by selective pressure and resource scarcity since different social and cultural conditions can increase or decrease them endogenously. The strong implication of the distinction between the three stages is that selective pressure should not change the social orientation of the firm. Whilst some sceptics of the importance of institutional variety and change see market pressure as a levelling force able to induce isomorphism even in the presence of different ownership and organizational forms, it has been shown that the processes leading to different distributive equilibria in for-profit firms and social cooperatives may differ markedly. The roots of these differences are found in different organizational routines even in the presence of market pressure. Whilst selection defines survival possibilities, the imitation of different routines serves for the coherent transmission of different organizational schemes and for cultural evolution. Hence the ability to reproduce particular routines defines social vocation of the organization, while selective pressure bears on the survival and expansion potential without changing the public benefit vocation.

4. CONCLUSION

This contribution has sought to shed new light on the nature of institutional variability, and on the process by which new organizational routines and governance arrangements are created in the context of the evolution of the social economy. The example used has been the emergence and diffusion of socially oriented entrepreneurial forms, which in Italy since the 1980s have been mainly represented by social cooperatives. The evolution of this new phenomenon can furnish new understanding of how individual aspirations and public policy can be conjugated to transform societal ideals into new entrepreneurial forms based on the involvement of different actors, and on the pursuit of social objectives other than maximization of the organization's economic value.

It has been shown that institutional creation and variability, far from being an isolated and marginal phenomenon, can indeed inform individual and collective action when framed by a suitable cultural and institutional context and by a correct understanding of the objectives and organizational models to be pursued.

NOTES

1. For an interesting discussion of the relationship between the theory and the history of the firm see Parker (1986).
2. The law on the 'community interest company' was enacted in the UK in 2005. In the same year the Italian Parliament passed the law on the *Impresa Sociale* (no. 118/2005) that was completed by decree no.155/2006.
3. Though the category of social enterprises is spreading in many industrialized countries, first and foremost the UK, and would be more suitable for international comparisons, it is too early to construct an historical evaluation of this new phenomenon.
4. Some of the most advanced studies in social psychology (Tyler and Blader, 2000) show that procedural fairness has both a personal and a routinized component. Whilst the former is interactional in character, the latter is linked to formal decision-making rules.
5. To simplify our argument, the different selective forces impinging on social cooperatives and public bodies will not be treated here.

REFERENCES

Ben-Ner, A. and L. Putterman (1999), 'Values and institutions in economic analysis', in A. Ben-Ner and L. Putterman (eds), *Economics, Values and Organisation*, Cambridge: Cambridge University Press, pp. 3–72.

Benz, M. (2005), 'Not for the profit, but for the satisfaction? Evidence on worker well-being in non-profit firms', *Kyklos*, **58** (2), 155–76.

Borzaga, C. and S. Depedri (2005), 'Interpersonal relations and job satisfaction: some empirical results in social and community care services', in: B. Gui and R. Sugden (eds), *Economics and Social Interaction: Accounting for Interpersonal Relations*, Cambridge: Cambridge University Press, pp. 132–53.

Borzaga, C. and A. Ianes (2006), *L'economia della solidarietà. Storia e prospettive della cooperazione sociale*, Rome: Donzelli.

Borzaga, C. and M. Musella (eds) (n.d.), *Produttività ed efficienza nelle organizzazioni non profit. Il ruolo dei lavoratori e delle relazioni di lavoro*, Trento: Edizioni 31.

Borzaga, C. and A. Santuari (2001), 'From traditional co-operatives to innovative social enterprises', in C. Borzaga and J. Defourny (eds), *The Emergence of Social Enterprise*, London: Routledge, pp. 166–81.

Borzaga, C. and E. Tortia (2006), 'Worker motivations, job satisfaction and loyalty in public and non-profit social services', *Nonprofit and Voluntary Sector Quarterly*, **35** (2), 225–48.

Borzaga, C. and E. Tortia (2007), 'Social economy organizations in the theory of the firm', in E. Clarence and A. Noya (eds), *The Social Economy. Building Inclusive Communities*, Paris: OECD Publishing, pp. 23–60.

Borzaga, C. and E. Tortia (forthcoming), 'The economics of social enterprises: towards a new comprehensive approach', in L. Becchetti and C. Borzaga (eds), *The Economics of Social Responsibility*, London: Routledge.
Chandler, A.D., Jr. (1977), *The Visible Hand: The Managerial Revolution in American Business*, Cambridge, MA: Harvard University Press.
Chandler, A.D., Jr. (1990), *Scale and Scope. The Dynamics of Industrial Capitalism*, Cambridge, MA: Harvard University Press.
Depredi, S. (2007) 'Le politiche nei confronti delle risorse umane: selezione, coinvolgimento, formazion', *Impresa Sociale*, **17**(3), 150–59.
Hodgson, G.M. (2006), *Economics in the Shadows of Darwin and Marx. Essays on Institutional and Evolutionary Themes*, Cheltenham, UK and Northampton, MA, USA: Edward Elgar.
Ianes, A. (2008), 'La cooperazione sociale come storia d'impresa', *Imprese e Storia*, **36** (2), (forthcoming).
ISTAT (2006), 'Le cooperative sociali in Italia. Anno 2003, Informazioni no. 30, Rome.
ISTAT (2007), 'Le cooperative sociali in Italia. Anno 2005', statistiche in breve, Rome.
Kreps, David (1990), 'Corporate culture and economic theory', in James Alt and Kenneth Shepsle (eds), *Perspectives on Positive Political Economy*, New York: Cambridge University Press, pp. 90–143.
Leete, L. (2000), 'Wage equity and employment motivation in nonprofit and for-profit organizations', *Journal of Economic Behavior and Organization*, **43** (4), 423–46.
Nelson, R. and S.G. Winter (1982), *An Evolutionary Theory of Economic Change*, Cambridge, MA: Belknap Press of Harvard University Press.
Parker, W.N. (ed.) (1986), *Economic History and the Modern Economist*, Oxford: Blackwell.
Polanyi, M. (1958), *Personal Knowledge: Towards a Post-critical Philosophy*, Chicago, IL: University of Chicago Press.
Sacchetti, S. and R. Sugden (forthcoming), 'The organization of production and its publics: mental proximity, markets and hierarchies', *Review of Social Economy*.
Sacchetti, S. and E.C. Tortia (2008), 'Dall' organizzazione multi-stakeholder all' impresa reticolare', *Impresa Sociale*, **18** (4), 104–24.
Scalvini, F. (1989), 'Organizzare imprese sociali per strategie di solidarietà', 2a Assemblea organizzativa Federsolidarietà, Castellamare di Stabia, 26-28 October, Rome, Federsolidarietà.
Tortia, E. (2008), Perceived fairness and worker well-being: survey-based findings from Italy', *Journal of Socio-Economics*, **37** (5), 2080–94.
Tortia, E. (forthcoming), 'The impact of social enterprises on output, employment and welfare', in L. Becchetti and C. Borzaga (eds), *The Economics of Social Responsibility*, London: Routledge.
Tyler, T.M. and S.L. Blader (2000), *Cooperation in Groups. Procedural Justice, Social Identity, and Behavioral Engagement*, Philadelphia, PA: Psychology Press.
Weisbrod, B.A. (1998), *To Profit or Not to Profit? The Commercial Transformation of the Non-Profit Sector*, New York: Cambridge University Press.

9. Creativity in economic development: space in an inferno

Silvia Sacchetti and Roger Sugden

> The inferno of the living is not something that will be; if there is one, it is what is already here, the inferno where we live every day, that we form by being together. There are two ways to escape suffering it. The first is easy for many: accept the inferno and become such a part of it that you can no longer see it. The second is risky and demands constant vigilance and apprehension; seek and learn to recognize who and what, in the midst of the inferno, are not inferno, then make them endure, give them space.
>
> Calvino, 1972, p. 164

1. INTRODUCTION

Previous analysis suggests links between the development of economies and the stimulation of people's creativity. Both inter- and intra-country variations in development (Henderson et al., 2001) are associated with such links. Over recent years these arguments have often been framed in terms of achieving regional 'competitiveness' in a global market economy (Bristow, 2005), for example through innovation in industries in general (Florida, 2002a) and through the success of creative (or cultural) industries in particular (Caves, 2000) . In this chapter, however, we examine the links from a different perspective. Having critically considered recent contributions on the impact of creativity based on accepted notions of competitiveness and prosperity, we offer a novel perspective that stresses a role for 'publics' in creatively shaping processes of economic development.

The association between creativity and innovation suggested by Florida (2002a) is that places with 'a high concentration of bohemians . . . reflect an underlying set of conditions or milieu which is open and attractive to talented and creative people of all sorts . . . and thus create a place-based environment that is conducive to the birth, growth and development of new and high-technology industries' (p. 68). His focus is essentially market orientated, capitalist success, and his analysis could comfortably fit into commonly made arguments about regional competitiveness (see also

Gordon and McCann (2005) on the geography of commercial innovation, creativity and entrepreneurship). Likewise could be said of many economic arguments about creative industries, which supply the 'goods and services that we broadly associate with cultural, artistic, or simply entertainment value. They include book and magazine publishing, the visual arts (painting, sculpture), the performing arts (theatre, opera, concerts, dance), sound recordings, cinema and TV films, even fashion, toys and games' (Caves, 2000, p. 1). The idea is that visual and performing arts, indeed cultural activities more generally, are associated with the production of goods and services that can be traded on markets to desirable effect for particular localities and for particular sets of people in terms of wealth, employment opportunities and so on. Consider, for example, Caves (2000) on the geography of creativity; Neff (2005) on the digital media industry in New York; Leslie and Rantisi (2006) on urban economic development and the interplay between 'economic' and 'cultural' factors in Montreal's design economy.

We would agree that such analysis has interest and relevance, but our focus is quite distinct. We concentrate on creativity amongst people in general, rather than Florida's concern with bohemian groups or indeed a 'creative class' (Florida, 2002b), and rather than certain sorts of industries. Artistic activities are reasoned to be of significance because of their potential to stimulate creativity across all sectors. Moreover, our special focus is not the attainment of competitiveness as commonly understood. We share the concerns of Bristow (2005) that, despite 'confusion as to what the concept actually means and how it can be effectively operationalised' (p. 286), competitiveness has become a theory-leading, hegemonic discourse in public policy circles, especially in so-called developed countries. However we go further, questioning not only the assumed aims of competitiveness but, more specifically, the process for choosing those aims and, correspondingly, for choosing the means by which they are pursued. This is a governance-centric perspective.

Wojcik (2006) argues that 'economic geographers often talk about corporate governance, without mentioning the term or referring to corporate governance research, despite potential benefits from doing so' (p. 640), but he goes on to assert that 'the time is ripe for economic geography research to examine corporate governance concepts and literature more explicitly' (pp. 640–1). We follow that assertion. Our analysis is grounded in an appreciation of large firm governance, albeit extending beyond such organisations and with a distinct emphasis on governance defined in terms of strategic choice (unlike in Wojcik (2006)). Compare as well, for example, Leslie and Rantisi (2006), who pay no explicit attention to strategic choice per se and for whom governance of cultural industries refers in particular to the role of government and the state, markets and hierarchies.

Zeitlin (1974) argues that the power to govern (in other words, to control) a large corporation equates to the power to make the strategic decisions that determine its broad direction; these include decisions about its relationships with other corporations, with governments and with employees, and about its geographical orientation. More recently, this analysis has been used as a foundation for the so-called strategic choice framework, deploying a governance lens to view the activities of transnational corporations, networks and other forms of economic organisation, and to view regional, national and indeed global economies.

The basis of this framework is a heterodox economic analysis of the theory of the firm (Cowling and Sugden, 1998a), of the development of economies (Sugden and Wilson, 2002) and of forms of globalisation of production (Sugden and Wilson, 2005). The analysis focuses most especially on the governance of the transnational corporation and its impact on contemporary economies (inter alia Hymer (1972) on uneven development; Cowling and Sugden (1998b) on strategic international trade; Cowling and Tomlinson (2000) on Japan); hence on learning from the experiences of successful agglomerations in the likes of the Third Italy so as to nurture multinational networking in the public interest (see, for example, Cowling and Sugden (1999) on multinational webs; Sacchetti and Sugden (2003) on network forms; Sacchetti (2004) on knowledge; Branston et al. (2006a) on the public interest).

Whilst the strategic choice framework offers a dynamic institutional perspective that rejects a particular stress on the neoclassicism at the heart of both regional science and 'new economic geography' (Boschma and Frenken, 2006), its foundations nevertheless reflect, and can be argued to contribute insight on, concerns at the core of economic geography. For example, Scott (2004) asserts that economic geography is especially focused on two areas of study: on the one hand, transnational corporations, globalisation, etc., and on the other hand, spatial agglomeration analysis, rooted in the apparent success of particular regions like the Third Italy. These are precisely the areas of study providing critical foundations to the strategic choice approach.

In this chapter we use such an approach to consider creativity in economic development. In doing so we would also stress that our general research preoccupation is not with disciplinary boundaries, hence not per se with any one notion of the likes of 'economics', 'geography' or 'economic geography'. It is with the scientific analysis of (aspects of) economies, our working definition of an economy being: a complex of people whose (interacting) relations, behaviour and actions have consequences for how production is organised, hence implications for the satisfaction of (human and other) interests.[1] There can be no doubt that the ideas relevant to understanding an economy so conceptualised encompass at their core, spatial factors, as both determinants and outcomes. They clearly also entail myriad complex factors, and a consequence of bounded rationality is that our research does not presume to be all encompassing. Instead, we seek a coherent perspective that offers significant insight.

The analysis proceeds as follows. Section 2 lays the foundations for our appreciation of creativity by considering in detail the strategic choice approach to economic development, competitiveness and globalisation, rooting analysis in understanding of the transnational corporation. This leads to an examination of the distinction between private and public interests, hence the possibility of the latter as a criterion for economic geography to assess realities. It concludes that public interests tend to be marginalised in people's typical experiences and, with that in mind, section 3 focuses on the kindling of people's creativity so that they might shape new strategic directions in the economies in which they have an interest. We advocate 'public creativity forums' and explore what that would mean. Section 3.1 discusses a notion of 'creative atmosphere', related to but distinct from Marshall's (1920) concept of industrial atmosphere. Section 3.2 considers visual and performing arts, music, cinema and indeed artistic activities more generally as a viaticum for the stimulation and expression of people's creativity, thus a potentially significant influence on strategic direction across all sectors. Section 4 offers concluding remarks: a summary, and a suggestion to consider new action research in economic geography.

2. POWER AND UNEVEN DEVELOPMENT

The concentration of power and uneven development are in many respects well recognised in economic geography. For example, Henderson et al. (2001) review analysis of uneven development across and within countries, and its relation to issues such as trade, investment, technology, urbanisation and income. For them, 'the most striking fact about the economic geography of the world is the uneven distribution of activity' (p. 81), reflected in 54 per cent of world GDP being produced by countries occupying 10 per cent of the land mass. Similarly Coe and Yeung (2001), asserting that not only is 'uneven development . . . the single most visible structural outcome of globalisation processes' (p. 370), it has been studied by radical geographers since well before globalisation became a key word in the social sciences in the 1990s. Moreover, they identify two elements to the unevenness, structural (different impacts across sectors in a given territory) and geographical (variations across territories), and relate the latter to 'uneven power relations underlying most global production chains such that some segments of these chains have disproportionately greater power and control over other segments' (p. 371). It is notable that this recognition of concentrated power applies not only to the power associated with particular regions, but also to that of particular firms. Consider for example, Fold (2001), highlighting the impacts of large producers in the chocolate industry in Europe on cocoa production in West Africa, and linking those with the influences of the structural adjustment programmes stimulated by the World Bank and International Monetary Fund (IMF).

What has tended to be ignored in these analyses, however, is a consideration of strategic choice as the source of power, hence as a root cause of uneven development.

The potential significance of this perspective was indicated in the heterodox economics literature by Hymer's (1972) seminal contribution – also well before globalisation became a popular concern in the social sciences. He recognised transnational corporations as likely to be especially influential organisations in the world economy, and contemplated what this would imply by extrapolating from an appreciation of their place in the historical development of US capitalism. Hymer argued (ibid., p. 50):

> One would expect to find the highest offices of the [transnational] corporations concentrated in the world's major cities . . . These . . . will be . . . major centres of high-level strategic planning. Lesser cities throughout the world will deal with the day-to-day operations of specific local problems. These in turn will be arranged in a hierarchical fashion: the larger and more important ones will contain regional corporate headquarters, while the smaller ones will be confined to lower level activities. Since business is usually the core of the city, geographical specialisation will come to reflect the hierarchy of corporate decision making, and the occupational distribution of labour in a city or region will depend upon its function in the international economic system.

Hymer's analysis has been criticised in its details because it simplifies a complex reality, yet it has also been argued on the basis of the empirical evidence that if it is accepted for the characterisation that it purports to be, then it offers insight (Dicken, 1992; Cowling and Sugden, 1994). Indeed, his analysis has received increased theoretical and empirical attention during the 1990s as well as at the start of the new millennium.

2.1 The Strategic Choice Framework

The focus on corporations and strategy is taken up in Cowling and Sugden (1998a), grounding analysis in Coase (1937, 1991) but critiquing mainstream economic theories, including the transactions cost approach rests on Williamson (1975). Accommodating debates about differences across corporations with their 'homes' in different countries, not least the idea of distinctions between Anglo-US and Japanese firms (Aoki, 1990), as well as debates about flexible specialisation – reorganisation by large corporations along lines implied by successful agglomerations of small firms in, for instance, the Third Italy (Sabel, 1988) – the strategy perspective reasons that large corporations are characterised by an essential asymmetry: a concentration in the power to make strategic decisions over the direction of production. Drawing on Zeitlin (1974), the basic idea is as follows (see also Branston et al., 2006a; Bailey et al., 2006):

- A transnational corporation can be shown to have an explicit and/or implicit strategy that is more or less coherent;
- This strategy encompasses the aims of the corporation, both what those aims are and the broad terms for their pursuit;
- The strategy is especially (albeit not all) important in determining the activity that the corporation undertakes;
- The strategy has determinants, including choices that can be conscious and/or unconscious;[2]
- The power to choose its strategy equates to the power to govern the corporation: to govern is to have the ability to choose – subject to constraints – both the aims of the corporation, and the broad terms for their pursuit;
- The power to govern typically lies with a subset of those with an interest in the corporation's activities, despite the objections and perhaps resistance of other interested parties.[3]

Sugden and Wilson (2002) apply this perspective to a consideration of the development of economies. They position Hymer's (1972) analysis of uneven development in the context of the agenda supplied by the 'Washington consensus' (Williamson, 1990; Rodrik, 1996), which places transnational corporations at its heart and a version of which has been a strong prevailing influence throughout most countries of the world since the early 1980s (as illustrated by Fold's (2001) aforementioned analysis of the World Bank and IMF backed structural adjustment programmes in West Africa). Illustrating from South Africa and Nicaragua, and as with transnational corporations, they reason that insofar as the aims of economic development for a particular region are chosen, the process is typically characterised by a concentration of power, with the institutions at the core of the Washington consensus being especially influential – for example through the World Bank's (1999) emphasis on GNP per capita, or the UNDP's (1997) broader approach based upon its Human

Development Index. This conclusion accords with Nelson Mandela's perception that 'people living in poverty have the least access to power to shape policies – to shape their future' (Mandela, 2006, p. 1).

The strategic choice framework recognises that, for any region, there are many people with an interest in its economic development, and many who might have a view on development aims (Branston et al., 2006b), it is just that in current practice they tend to have little or no effective voice. They would include those who currently live in the region, as well as those who might live there in the future, not least potential immigrants. Moreover, the development in and around the region would likely impact on, and be impacted by, development elsewhere – in other places in the same country, continent and indeed the world. People in those places might have interests that are relevant, and possibly experiences which they could exchange with others, so that together people and regions might all find more desirable development aims.

Analogous arguments to those about economic development are also made by Sugden and Wilson (2005) when analysing the conceptualisation of globalisation. They suggest that models of development correspond to models of globalisation. For example, the Washington consensus development agenda is associated with a Washington consensus form of globalisation; the aims of both are identical, and each implies a parallel set of strategic choices to the other. This reasoning overlaps with that in Coe and Yeung (2001), who stress that 'economic globalisation is not some kind of immutable inevitability, but a set of processes that is socially constructed, and therefore can be encouraged or resisted by actors/institutions at various scales' (p. 368). In other words, we might view strategic decisions to pursue a Washington consensus development agenda as paralleled by strategic decisions to pursue a Washington consensus form of globalisation.

This perspective can also be extended to a consideration of competitiveness. Consistent with comments in this chapter's Introduction, competitiveness is a conveniently flexible and loosely used concept; as Poerksen (1995) said of 'development', and as we might observe of 'globalisation', 'competitiveness' is a plastic word. Nonetheless, it is interesting to note Bristow's (2005) observation that 'the regional competitiveness discourse ignores the possibility that regional prosperity might be achieved . . . by the development of community or social enterprises which meet broader social and environmental . . . objectives. As a consequence, policies tend to prioritise rather narrow, private sector originated agendas at the expense of broader regeneration initiatives' (p. 295). That is to say, the aims of regional competitiveness are confined, provided by the private sector agendas that inform, and are therefore in line with, the Washington consensus development agenda which seeks, for example, to enable private enterprise and in particular transnational corporations to freely move goods, services and capital across economies.[4]

2.2 The Interests of Publics

Bristow's (2005) recognition that private agendas occupy centre stage can be re-interpreted as public interests being confined to the margins, causing us to raise the possibility that the interests of publics might provide a suitable evaluation criterion for economic geography. This follows Long (1990), who proposes the public interest as a criterion for research and policy in public administration and political science, and Branston et al. (2006a), who suggest it for much of economics.

According to Dewey's ([1927] 1988) seminal work in political and social philosophy, an action – such as making a strategic choice – might have significant consequences for two categories of people: private interests, those who are directly engaged in the action; public interests, those not directly engaged (see also Young, 2002). An action might be associated with multiple private interests and multiple publics. Each public is seen to have shared concerns.

Drawing on Dewey ([1927] 1988), Long (1990) views a public interest as an evolving consensus, a criterion agreed upon by a public and against which private actions can be assessed. For him, therefore, the 'consequences of private parties' actions create a public'. This occurs as a process of discovery through which the public identifies a shared concern with the effects of private actions, as well as the need for controlling such results. It follows that 'the public's shared concern with consequences is a public interest' (p. 171). Referring to this, Branston et al. (2006a, 195) identify 'the public interest in a corporation's activities in general and in its strategies in particular as the agreed upon, evolving concerns amongst all of those indirectly and significantly affected by those activities and strategies (wherever they live, whatever their nationality)'.

To illustrate, according to the strategic choice framework, to the extent that the aims of the typical transnational corporation, and the broad terms for their pursuit, are chosen, the decision is made by a subset of those with an interest in the corporation's activities. That choice by private interests impacts on others, on publics. Positive outcomes discussed in the literature include effects on technological transfer and contingent employment growth, commonly argued as potentially desirable consequences of incoming foreign direct investment with respect to the development of localities (see, for instance, the appraisal of transnationals' impacts in Dicken (2007)). However, even in these cases we would argue that an exclusion issue remains, and that technological transfer and employment growth induced by transnationals have their shortcomings (Blomström, 1986; Blomström and Kokko, 2002).

Consider also, for example, the implications for international trade. Cowling and Sugden (1998b) suggest that 'free international trade' implies the freedom of the private interests governing transnational corporations to manage trade in pursuit of their own interests, despite the possibly adverse impacts on others. This includes, for example, managing trade in pursuit of a divide and rule approach to labour. The idea is that the strategic decision-makers of a transnational corporation might be concerned to improve their bargaining power with respect to employees, so as to improve profits. Accordingly, a corporation supplying markets across Europe might deliberately opt to produce the same goods in various countries, so that if employee industrial action in one country interrupts supply, that might be compensated by an increase in supply from elsewhere (on the basis that collective action tends to be more problematic for employees across rather than within countries). Such strategies clearly have consequences beyond the private interests making the choice; not least, the affected employees are a public with an interest in the action.

Similar arguments could be made in the analysis of uneven development, globalisation and regional competitiveness. Following Hymer (1972), concentrations in the power to govern corporations have significant effects on levels of development, wealth and poverty; those in poverty in so-called less developed countries have public interests in the strategic choices of transnationals. Sacchetti (2004), for instance, applies Hymer's divide and rule strategy (Hymer, 1972) to knowledge production and diffusion across countries. Referring to the international division of labour, and critical towards current faith in technological transfer, she argues – building on Marglin (1974) – that the geographical scattering of different activities, which follows strategic decisions taken by restricted groups organising activities transnationally, may jeopardise peoples' knowledge in those localities where concentration of operational and repetitive tasks occurs. Vicious cycles, as path dependence theories would explain (Nelson, 1994), might then start to build up, affecting institutions, for instance in the education system, by shaping strategies in ways that suit the transnational production system, possibly disregarding the interests of different publics.

Likewise, by analysing globalisation, we can identify issues of concentration of strategic decision-making power in a Washington consensus stimulated reality (Sugden and Wilson, 2005). This suggests the existence of publics with interest that are not being met, as reflected in the frustrations and actions of so-called anti-globalisation movements. These are made up of diverse people and groups, most of whom are probably not against globalisation in the sense of using new technologies and opportunities to decrease the territorial barriers between people (Sugden and Wilson, 2005). They form interested publics, expressing their interests in protests against the outcomes of current forms of globalisation, and against the ways in which those outcomes are being pursued.

In principle, a fundamental issue might be that public interests are being deliberately flouted, but even in the 1920s, Dewey ([1927] 1988, p. 314) identified another possibility, one that technological changes and the so-called new economy might make even more pertinent today (evidence of vociferous portions of anti-globalisation movements notwithstanding):

> Indirect, extensive, enduring and serious consequences of conjoint and interacting behaviour call a public into existence having a common interest in controlling . . . consequences. But the machine age has so enormously expanded, multiplied, intensified and complicated the scope of the indirect consequences . . . that the resultant public cannot identify and distinguish itself.

He sees a special problem with 'the eclipse of the public' (p. 304), which 'seems to be lost' (p. 308), 'amorphous and unarticulated' (p. 317). For Dewey (ibid., p. 327), 'the prime difficulty' for acting in the public interest is discovery of 'the means by which a scattered, mobile and manifold public may so recognise itself as to define and express its interests.'

3. CREATIVITY, COMMUNICATION AND PUBLIC SPACE

An implication of our analysis of power, uneven development and strategic choice is that confining the interests of publics to the margins raises fundamental queries about the exercise of creativity in economic development. More specifically, excluding actual and potential publics from strategic choice processes would seem to deny the people who make up those publics the opportunity to develop and use their imagination and ideas (their creativity) in the shaping and determination of economic strategy. For example, echoing the words of Bristow (2005), prioritising narrow, private inputs in the regional competitiveness discourse ignores the possibility that regional economic prosperity might be achieved by the development of innovative economic strategies that are stimulated by the imagination and ideas of currently excluded people, who might also catalyse the targeting of broader and even currently unimagined aims. This would have no import if the currently excluded people have no inherent creativity to bring to bear, but that seems most unlikely. Consider, for instance, the thoughts of Chomsky (1975) on the education of children.

He argues that each person has an intrinsic, unique creativity and that this needs to be nurtured, hence he advocates education aimed 'to provide the soil and the freedom required for the growth of this creative impulse' (p. 164).

Moreover, an exclusion of publics might be associated with a downward spiral: people's creativity is not being exercised, thus not stimulated, explored and enhanced; therefore their capabilities to exercise imagination are truncated and even lost; therefore their creativity is not exercised . . . This might lead to, and be fed by, perceptions of 'not counting'. Dewey's (1927) focus on publics being eclipsed is also a relevant factor: perhaps a reason for the eclipse is an exclusion which, over time, becomes self reinforcing, resulting in a public losing sight of itself, of not even being aware of its own existence.

Viewed from the opposite direction, however, this analysis implies a challenge and potential opportunity: people in actual and potential publics might seek to kindle their imagination and ideas, to exercise their creativity, thereby attempt to seize opportunities to shape and determine strategic choices influencing the development of the economies in which they have an interest. Although the precise consequences that this might have are unclear, we would hypothesise that there would be opportunities to pursue new avenues of economic prosperity, simply because more people would be exercising their creativity and would be doing so in search of new strategies (Sugden and Wilson, 2005).

As for how to enable creative publics, a first step is suggested by Dewey's ([1927] 1988) consideration of the means by which lost publics might find themselves. For him, 'the essential need . . . is the improvement of the methods and conditions of debate, discussion and persuasion. That is *the* problem of the public' (p. 365). The necessary continuous, inclusive discourse is argued to be in part an attitude acquired by nurtured habit, and he stresses knowledge, learning and communication:

> An obvious requirement is freedom of social inquiry and of distribution of its conclusions. . . . There can be no public without full publicity in respect to all consequences which concern it. Whatever obstructs and restricts publicity, limits and distorts public opinion and checks and distorts thinking on social affairs. Without freedom of expression, not even methods of social inquiry can be developed. For tools can be evolved and perfected only in operation; in application to observing, reporting and organizing actual subject-matter; and this application *cannot occur save through free and systematic communication.* (ibid., emphasis added, pp. 339–40)

A related stress on communication is also seen in analysis of the competence-based view of the knowledge economy (reviewed in the context of economic geography by Gertler, 2001). For example, Amin and Cohendet (2000, p. 99) consider effective knowledge circulation in an organisation as associated with 'dialogue, discussion, experience-sharing', and to 'socialising activities'. At issue are cognitive phenomena generated through interaction. There is a particular focus on 'relationships, based on shared norms and conventions' and on communities of practice, 'groups of individuals informally bound together by shared expertise and a common problem' (Gertler, 2001, p. 18). The reference to a common problem echoes the common interest essential to a public, and suggests that the identification of publics might learn from analysis of communities of practice and the competence-based view of the knowledge economy more generally.

Accordingly we infer that creative publics might be enabled, in the first instance, by the construction and nurturing of 'public creativity forums', spaces where people – the members of actual and potential publics – can freely engage with each other in learning, discussion and debate about the development of the economies in which they have an interest; where people's relations are characterised by shared values of openness, of their essence rejecting any significant influence of private over public interests, so as to avoid outcomes that are essentially similar to the current realities of concentrated power in economic development, competitiveness and globalisation; where people recognise and cultivate a concern with each other's ideas and perspectives through reasoned and coherent understanding, so as to anchor the foundations for the interest of each public in rational argument and analysis. (Using the terminology of Scott (2006, p. 3), a public creativity forum can be viewed as a specific type of 'creative field', a notion that 'can be used to describe any system of social relationships that shapes or influences human ingenuity and inventiveness and that is the site of concomitant innovations'.)

We hypothesise that with public creativity forums as a basis, people could start – with respect for each other and hence for publics – to discuss and talk with others, to share arguments and mutually influence ideas by increasing – through communication – the diversity of perspectives and possibilities on the strategic choices that underpin the development of economies.

3.1 Creative Atmosphere

It follows from our analysis thus far that public creativity forums would have an atmosphere in some ways similar to the 'industrial atmosphere' that Marshall (1920) identified as characteristic of certain places. He refers to people in an agglomerated industry receiving 'advantages . . . from near neighbourhood to one another. The mysteries of the trade become no mysteries; but are as it were in the air, and children learn them unconsciously' (p. 271). For example, 'if one man starts a new idea, it is taken up by others and combined with suggestions of their own; and thus it becomes the source of further new ideas' (ibid.). Public creativity forums would be similarly spaces where the exercise of imagination and the pursuit of ideas are in the atmosphere; spaces where ideas flow between people, learning from each other, shaping each other's perspectives.

However, a crucial difference is that when analysing industrial atmosphere, Marshall (1920) is not especially concerned with strategic choices in an economy. Furthermore, whilst he considers place, we focus on the more general notion of creative atmosphere conceived in socio-economic space.

In this respect our argument follows the likes of Lorentzen (2007) (see also, for example, Agrawal et al. (2006), who use empirical evidence on patenting to consider the significance of social relationships in altering the impact of geographical proximity on knowledge flows; the discussions of relational proximity in Amin and Cohendet (2000) and Gertler (2001); and Boschma's (2005) consideration of proximity concepts more broadly). Lorentzen refers to an agreement in the literature that knowledge is developed and exchanged in social spaces, but she criticises the tendency in research on regional development policy to degenerate this insight into territorial determinism; analysis tends to focus on place (industrial districts, milieus . . .) rather than space, when it is the latter that is most relevant to knowledge flows and innovation.

The implication we draw from Lorentzen is that creative atmospheres can be generated and renewed through multidimensional spaces. In some circumstances these might include a special territorial dimension – a public creativity forum rooted in and developed from a particular region is certainly conceivable – but not necessarily. More generally forums might develop in different, inter-acting and overlapping scales – for example in creativity festivals, conferences, meetings, projects (including university-linked projects), both within and across territories, international and local (see also Dicken et al. (2001) on multiple scales in the global economy).

3.2 Artistic Activities

It was observed in the Introduction that the subject of creative industries – hence visual and performing arts, music, cinema and indeed artistic activities more generally – has become topical in large part because of their potential for contributing to wealth creation in a competitive market environment. However, an implication of our analysis is another, quite distinct explanation for a telling impact: because artistic activities are a viaticum for the stimulation and expression of people's creativity,[5] they are a potentially significant stimulant in the construction and development of public creativity forums. It can even be hypothesised that people's openness and access to artistic activities is a crucible for evolving public creativity forums.

This direction of causality accords with Scott (2004, 488): writing of the recent cultural turn in economic geography, he identifies 'a growing conviction that not only were certain earlier generations of geographers and other social scientists incorrect to regard culture simply as an outcome of underlying economic realities, but that these realities themselves are in fundamental ways subject to the play of cultural forces'. This is also a point long before recognised but since lost:

> Adam Smith, the master builder of models in both economics and ethics, was . . . as thoroughly comfortable drawing his lessons from Hamlet as from Hume. Like the creator of a patchwork quilt, he dapples [sic] in dramas, dabs in novels, dusts in some poetry and bellows opera. *It is not simply that Smith likes and employs the arts. Rather . . . Smith finds the arts essential for the task at hand – understanding and moulding human conscience.* (Wight, 2006, p. 156, emphasis added)

Moreover, in urging the significance of artistic activities for the construction and nurturing of public creativity forums the intention is not to reduce art to an instrument of economic development, which compares starkly with what Sir John Tusa (former managing director of London's Barbican Centre) sees as the approach of Tony Blair's UK government: 'what they have insisted is that the arts must fulfil a social, political, environmental, educational or economic purpose – in other words they must be an "instrument" for "delivering" other government policies. The impact on some museums and galleries, according to one observer, is that "scholarship, collection and curating are out of the window – the new breed of manager/directors is interested only in cramming into their buildings as many schoolchildren as possible"' (Tusa, 2007, p. 11). On the contrary, although recognising that creative activities can have *ex post* consequences, we see neither art nor artists as an *ex ante* instrument for achieving any particular goals, instead hypothesising that the stimulation of people's creativity in the economic sphere is linked in a holistic sense to the freedom of artists to express themselves in whatever directions they see fit. In part, the underlying intuition is that the link between economic creativity and artistic expression is simply that the latter, of its essence, without recourse to plans or instruments, provides a direct stimulant for activity in other areas, including in thinking about strategic choices for the development of an economy. In part it is the sense that only in an environment – a creative atmosphere – of artistic freedom can people be emancipated to realise the full potential of their creativity in the economic sphere; any attempt to plan *ex ante* functional consequences might limit the achievements of artistic activities, and in the extreme any restraint on artistic freedom risks the constraining of imagination and analytical powers more generally, including in the economy.[6]

4. CONCLUDING REMARKS: NEW DIRECTIONS FOR ECONOMIC GEOGRAPHY

This chapter offers a distinct perspective on the links between the development of economies and the stimulation of people's creativity. It emphasises strategic choice as a source of power, hence as a determinant of uneven development. The ideas are explored through a consideration of the nature of the transnational corporation, the development of economies, globalisation and regional competitiveness. Private interests are observed to occupy centre stage in the realities people typically experience, and we suggest the possibility that the interests of publics might provide an insightful evaluation criterion for economic geography. The marginalisation of publics is linked to their not being aware of their own existence, and from this we identify the prospect of people in actual and potential publics kindling their imagination and ideas so as to shape new strategic directions in the economies in which they have an interest.

Specifically, the construction and development of public creativity forums are advocated as an initial step in a possible alteration in strategic choice processes, perhaps moving current economic development and globalisation processes from a Washington-consensus-based focus on narrow interests, hence uneven development, towards alternatives that break the constraints implied by typical approaches to regional competitiveness. These public creativity forums are viewed as spaces defined not in physical terms but according to the embracing of certain types of relations, namely those aimed at free communication about strategic choices on the development of economies and based upon shared values: openness; a rejection of private interests dominating the interests of publics; people's concern, through reasoned and coherent understanding, with each other's ideas and perspectives. Forums are described as having creative atmospheres in multidimensional spaces; they might develop in varied interacting and overlapping scales both within and across international and local territories. Echoing the words of Calvino (1972) with which the chapter is introduced, they might provide spaces for people to step outside the economic inferno that most experience as a consequence of the ignoring of the interest of publics.

The chapter identifies visual and performing arts, music, cinema and indeed artistic activities more generally as a viaticum for the stimulation and expression of people's creativity, thus a potentially significant influence on the construction and development of public creativity forums. This is an emphasis on artistic activities that differs markedly from the preoccupations of much other literature on creativity, certainly in economics, where analysis of creative industries tends to concentrate on a competition amongst peoples to produce outcomes that can be transacted on a market. In contrast, public creativity forums are concerned with the development and application of each person's creativity, whether or not this can be displayed and realised through goods and services that can be transacted on the market. From this perspective the significance of artistic activities is their stimulating affects on people, hence publics, with interests in any sector ('creative industry' or otherwise).

We would suggest that this analysis implies a new avenue for public policy: to provide supporting instruments for the development of public creativity forums. Included in this there is a clear opportunity for regional policy, for towns and localities to foster the emergence of forums related to the economies in which their citizens have an interest.[7]

Moreover, because they have mutual learning, discussion and debate at their heart, another significant catalyst in forum formation and operation would be research and learning activities.

Information and knowledge would be crucial as both inputs and outputs to public creativity forums, perhaps suggesting that there is a sense in which any education system might provide suitable catalytic effects. However, the implications of Chomsky's (1975) perspective are that something more particular would be ideal. We referred earlier to his comments about people's intrinsic creativity. From that basis he argues that the purpose of education is not 'to control' a person's:

> growth to a specific, predetermined end, because any such end must be established by arbitrary authoritarian means; rather, the purpose of education must be to permit the growing principle of life to take its own individual course, and to facilitate this process by sympathy, encouragement, and challenge, and by developing a rich and differentiated context and environment (p. 164).

In other words, Chomsky appears to reject the concentration of power in the governance of people's creative potential as a necessary consequence of each person having – and being able to develop – their own intrinsic creativity. Accordingly, we infer a correspondence between on the one hand, public creativity forums grounded on Chomsky's analysis and, on the other hand, education processes that similarly nurture and encourage people's intrinsic creativity.

A specific dimension of these education processes would be universities aimed at providing research and learning activities on such a basis, in particular without concentration in the power to determine their strategic direction (on which see Sugden's (2004) application of the strategic choice framework, rejecting an approach to the organisation of universities which mimics transnational corporations). This has implications not only for how each university is governed, but also for how universities relate to each other and for their regional spread (on which see Andersson et al. (2004), discussing Sweden's policy to decentralise higher education).

Furthermore, our analysis points to the desirability of new studies in economic geography. For example, research to show more precisely what would be entailed in shifting the interests of publics to centre stage, not least in the context of particular cases. Likewise the detailed effects of such a shift; it is one thing to reason that concentrated power in strategic decision-making is associated with uneven development and a constraining of people's creative capacities (as has been done in this chapter), it is another to present scientific empirical evidence on the hypothesis that unleashing wider creativity through enabling publics would open new opportunities to pursue new avenues of economic prosperity. There is also a pressing need for cooperation across researchers with particular expertise in economic geography and those with particular knowledge about artistic activities, with the objective of better understanding the stimulatory effects of such activities in the economic sphere.

We envisage research on, for example, the economy of particular places (urban and non-urban) positioned in their broader spatial context, so as to identify actual and potential publics with interests in the development of the economy; to study those interests – their formation, expression and influence – in their different, interacting and overlapping scales; to examine creativity in those publics and, included in that, consider ways in which that creativity is and might be stimulated through artistic activities so as to impact on the strategic direction of economic activity.

Perhaps most significantly, however, we urge researchers in economic geography to consider engaging in embryonic public creativity spaces, and indeed contributing to their being conceived; to think about designing and undertaking their work in active attempts to catalyse the development of such spaces.[8]

ACKNOWLEDGEMENTS

The authors would like to thank James R Wilson for comments and discussion on various aspects of the analysis. Thanks also to participants at the Ideas Laboratory of the International Festival on Creativity and Economic Development (Gambettola, Italy, May 2007), where parts of the chapter were discussed.

NOTES

1. The stress on people's relations behaviour and actions is rooted in the seminal views of Marshall (1920) and Robbins (1932), and the focus on production and organisation is in line with Backhouse (2002). We would also highlight that the definition is relatively open in terms of what people might seek from an economy, encapsulating material and non-material dimensions; and it removes Marshall's centre-stage concern with mankind in favour of considering human *and other* interests.
2. This perspective has strong ties with Penrose (1952, p. 818): 'there is considerable evidence that . . . many decisions are reached after a conscious consideration of alternatives, and that men have a wide range of genuine choices'.
3. The applicability of this perspective across countries and legal jurisdictions is implicitly addressed in a growing literature on convergence in corporate governance. See for example, Wojcik (2006), examining practice across Europe and finding evidence of convergence to an Anglo-US model.
4. Compare Branston et al. (2006b), offering the prospect of a conceptualisation of competitiveness that is much broader, albeit not arguing that broad approach is currently pursued in practice.
5. Support for this assertion might come from artists themselves: inter alia, for Wordsworth (1802; quoted in Knowles, 1999, 832) 'poetry is the breath and finer spirit of all knowledge'; for Cartier-Bresson (1952; quoted in Knowles, 1999, 193) 'photography is the simultaneous recognition, in a fraction of a second, of the significance of an event as well as of a precise organisation of forms which give that event its proper expression'; for de Mille (1975; quoted in Knowles, 1999, 257) 'the truest expression of a people is in its dances and its music'.
6. Having recognised this, however, we would not suggest that it is necessarily desirable to free artistic activities from all and any ethical constraints. In particular, it might be argued that human and other species have inalienable rights.
7. Policy support at national and supra-national levels might be considered a direct confrontation to the extant powers of transnational corporations and other organisations, such as the World Bank and IMF, which are currently so influential in setting agendas for development, globalisation and competitiveness. Hence it might be especially prone to undermining from that power, implying that subnational levels might be deemed particularly appropriate spheres for policy action.
8. On the relation between science and society, and most notably on aspects of what sustained interaction would entail, see the discussion of regional science in Barnes (2004), whose approach has overlaps with the more general analysis of networks in Dicken et al. (2001).

REFERENCES

Agrawal, A., I. Cockburn and J. McHale (2006), 'Gone but not forgotten: knowledge flows, labor mobility and enduring social relationships', *Journal of Economic Geography*, **6**, 571–91.

Amin, A. and P. Cohendet (2000), 'Organizational learning and governance through embedded practices', *Journal of Management and Governance*, **4**, 93–116.

Amin, Ash and Patrick Cohendet (2005), 'Geographies of knowledge formation in firms', *Industry and Innovation*, **12** (4), 465–86.
Andersson, R., J.M. Quigley and M. Wilhelmson (2004), 'University decentralization as a regional policy: the Swedish experiment', *Journal of Economic Geography*, **4**, 371–88.
Aoki, M. (1990), 'Toward an economic model of the Japanese firm', *Journal of Economic Literature*, **XXIIX**, 1–27.
Backhouse, R.E. (2002), *The Penguin History of Economics*, London: Penguin.
Bailey, D., L. De Propris, R. Sugden and J.R. Wilson (2006), 'Public policy for economic competitiveness: an analytical framework and a research agenda', *International Review of Applied Economics*, **20** (5), 555–72.
Barnes, Trevor J. (2004), 'The rise (and decline) of American regional science: lessons from the new economic geography?', *Journal of Economic Geography*, **4** (2), 107–29.
Blomström, M. (1986), 'Foreign investment and productive efficiency: the case of Mexico', *Journal of Industrial Economics*, **35** (1), 97–110.
Blomström, M. and A. Kokko (2002), 'FDI and human capital: a research agenda', OECD technical papers no. 02/195, Paris.
Boschma, R.A. (2005), 'Proximity and innovation: a critical assessment', *Regional Studies*, **39**, 61–74.
Boschma, R.A. and C. Frenken (2006), 'Why is economic geography not an evolutionary science? Towards an evolutionary economic geography', *Journal of Economic Geography*, **6**, 273–302.
Branston, J.R., K. Cowling and R. Sugden (2006a), 'Corporate governance and the public interest', *International Review of Applied Economics*, **20** (2), 189–212.
Branston, J.R., L. Rubini, R. Sugden and J.R. Wilson (2006b), 'The healthy development of economies: a strategic framework for competitiveness in the health industry', *Review of Social Economy*, **LXIV** (3), 301–29.
Bristow, G. (2005), 'Everyone's a "winner": problematising the discourse of regional competitiveness', *Journal of Economic Geography*, **5**, 285–305.
Calvino, I. ([1972] 1993) *Le Città Invisibili*, originally published in 1972, Milan: Oscar Mondadori, Turin: Ginaude.
Cartier-Bresson, H. (1952), *The Decisive Moment*, as quoted in E. Knowles (ed.) (1999), *The Oxford Dictionary of Quotations*, Oxford: Oxford University Press, p. 193.
Caves, R.E. (2000), *Creative Industries, Contracts between Art and Commerce*, Cambridge, MA: Harvard University Press.
Chomsky, N. (1975), 'Toward a humanistic conception of education', in W. Feinberg and H. Rosemont (eds), *Work, Technology and Education: Dissenting Essays in the Intellectual Foundations of American Educations*, Urban, IL: University of Illinois Press.
Chomsky, N. (1994), 'Prospects for democracy', talk given at MIT, Cambridge, MA, reproduced in C.P. Otero (ed.) (2003), *Chomsky on Democracy and Education*, London: RoutledgeFalmer.
Coase, R.H. (1937), 'The nature of the firm', *Economica*, **4**, 386–405.
Coase, R.H. (1991), 'The nature of the firm: meaning', in O.E. Williamson and S.G. Winter (eds), *The Nature of the Firm. Origins, Evolution and Development*, Oxford: Oxford University Press, pp. 48–60.
Coe, N.M. and H.W. Yeung (2001), 'Geographical perspectives on mapping globalisation', *Journal of Economic Geography*, **1**, 367–80.

Cowling, K. and R. Sugden (1994), *Beyond Capitalism: Towards a New World Economic Order*, London: Pinter.
Cowling, K. and R. Sugden (1998a), 'The essence of the modern corporation: markets, strategic decision-making and the theory of the firm', *Manchester School*, **66** (1), 59–86.
Cowling, K. and R. Sugden (1998b), 'Strategic trade policy reconsidered: national rivalry vs free trade vs international cooperation', *Kyklos*, **51**, 339–57.
Cowling, K. and R. Sugden (1999), 'The wealth of localities, regions and nations; developing multinational economies', *New Political Economy*, **4** (3), 361–78.
Cowling, K. and P.R. Tomlinson (2000), 'The Japanese crisis – a case of strategic failure', *Economic Journal*, **110** (464), F358–F381.
de Mille, A. (1975), *New York Times Magazine*, 11 May, as quoted in E. Knowles (ed.) (1999), *The Oxford Dictionary of Quotations*, Oxford: Oxford University Press, p. 257.
Dewey, J. ([1927] 1988), 'The public and its problems', in J.A. Boydston, *John Dewey. The Later Works Volume 2: 1925-1927*, Carbondale, IL and Edwardsville, IL: Southern Illinois University Press.
Dicken, P. (1992), *Global Shift: The Internationalization of Economic Activity*, London: Paul Chapman.
Dicken, P. (2007), *Global Shift: Mapping the Changing Contours of the World Economies*, London: Sage.
Dicken, P., P.F. Kelly, K. Olds and H.W. Yeung (2001), 'Chains and networks, territories and scales: towards a relational framework for analysing the global economy', *Global Networks*, **1** (2), 89–112.
Florida, R. (2002a), 'Bohemia and economic geography', *Journal of Economic Geography*, **2**, 55–71.
Florida, R. (2002b), *The Rise of the Creative Class*, New York: Basic Books.
Fold, N. (2001), 'Restructuring of the European chocolate industry and its impact on cocoa production in West Africa', *Journal of Economic Geography*, **1**, 405–20.
Gertler, Meric S. (2001), 'Best practice? Geography, learning and the institutional limits to strong convergence', *Journal of Economic Geography*, **1**, 5–26.
Gordon, I.R. and P. McCann (2005), 'Innovation, agglomeration, and regional development', *Journal of Economic Geography*, **5**, 523–43.
Henderson, J.V., Z. Shalizi and A.J. Venables (2001), 'Geography and development', *Journal of Economic Geography*, **1**, 81–105.
Hymer, S.H. (1972), 'The multinational corporation and the law of uneven development', in J.N. Bhagwati (ed.), *Economics and World Order: From the 1970s to the 1990s*, London: Macmillan; page numbers in the text refer to the reproduction in H. Radice (ed.) (1975), *International Firms and Modern Imperialism*, London: Penguin.
Knowles, E. (ed.) (1999), *The Oxford Dictionary of Quotations*, Oxford: Oxford University Press.
Leslie, D. and N.M. Rantisi (2006), 'Governing the design economy in Montreal, Canada', *Urban Affairs Review*, **41** (3), 309–37.
Long, N.E. (1990), 'Conceptual notes on the public interest for public administration and policy analysts', *Administration Society*, **22**, 170–81.
Lorentzen, A. (2007), 'Knowledge networks in local and global space', *Entrepreneurship and Regional Development*, **20** (6), 533–45.
Mandela, N. (2006), speech on receiving the Ambassador of Conscience

Award, accessed 13 March 2007 at www.artforamnesty.org/aoc/downloads/NelsonMandela_AOC2006.pdf.
Marglin, S. (1974), 'What do bosses do? The origins and functions of hierarchy in capitalist production', *Review of Radical Political Economics*, **6** (2), 60–120.
Marshall, A. (1920), *Principles of Economics*, London: Macmillan.
Neff, G. (2005), 'The changing place of cultural production: the location of social networks in a digital media industry', *Annals of the American Academy of Political and Social Science*, **597**, 134–52.
Nelson, R. (1994), 'The co-evolution of technology, industrial structure, and supporting institutions', *Industrial and Corporate* Change, **3** (1), 47–63, reprinted in G. Dosi, D. Teece and J. Chytry (eds) (1998), *Technology, Organization, and Competitiveness*, Oxford: Oxford University Press.
Penrose, E.T. (1952), 'Biological analogies in the theory of the firm', *American Economic Review*, **42** (5), 804–19.
Poerksen, U. (1995), *Plastic Words: The Tyranny of a Modular Language*, University Park, PA: Pennsylvania University Press.
Robbins, L.C. (1932), *An Essay on the Nature and Significance of Economic Science*, London: Macmillan.
Rodrik, D. (1996), 'Understanding economic policy reform', *Journal of Economic Literature*, **XXXIV**, 9–41.
Sabel, C. (1988), 'The re-emergence of regional economies', *Papers de Seminiari*, **29-30**, 71–140.
Sacchetti, S. (2004), 'Knowledge caps in industrial development', *New Political Economy*, **9** (3), 389–412.
Sacchetti, S. and R. Sugden (2003), 'The governance of networks and economic power: the nature and impact of subcontracting networks', *Journal of Economic Surveys*, **17** (5), 669–91.
Scott, A.J. (2004), 'A perspective of economic geography', *Journal of Economic Geography*, **4**, 479–99.
Scott, A.J. (2006), 'Entrepreneurship, innovation and industrial development: geography and the creative field revisited', *Small Business Economics*, **26**, 1–24.
Sugden, R. (2004), 'A small firm approach to the internationalisation of universities: a multinational perspective', *Higher Education Quarterly*, **58** (2/3), 114–35.
Sugden, R. and J.R. Wilson (2002), 'Economic development in the shadow of the consensus: a strategic decision-making approach', *Contributions to Political Economy*, **21**, 111–34.
Sugden, R. and J.R. Wilson (2005), 'Economic globalisation: dialectics, conceptualisation and choice', *Contributions to Political Economy*, **24**, 1–20.
Tusa, Sir J. (2007), 'Blair's lips may say "culture" . . . but his heart is not in art', *Observer Review*, 11 March, p. 11.
UNDP (1997), *Human Development Report*, Oxford: Oxford University Press.
Wight, J.B. (2006), 'Adam Smith's ethics and the "noble arts"', *Review of Social Economy*, **LXIV** (2), 155–80.
Williamson, J. (1990), 'The progress of policy reform in Latin America', in J. Williamson (ed.), *Latin American Adjustment: How Much has Happened*, Washington, DC: Institute for International Economics.
Williamson, O.E. (1975), *Markets and Hierarchies: Analysis and Antitrust Implications*, New York: Free Press.
Wojcik, D. (2006), 'Convergence in corporate governance: evidence from Europe

and the challenge for economic geography', *Journal of Economic Geography*, **6**, 639–60.
Wordsworth, W. (1802), *Lyrical Ballads*, quoted in E. Knowles (ed.) (1999), *The Oxford Dictionary of Quotations*, Oxford: Oxford University Press, p. 832.
World Bank (1999), *World Development Report 1999/2000*, Oxford: Oxford University Press.
Young, I.M. (2002), *Inclusion and Democracy*, Oxford: Oxford University Press.
Zeitlin, M. (1974), 'Corporate ownership and control: the large corporation and the capitalist class', *American Journal of Sociology*, **79** (5), 1073–119.

PART IV

Creative activities: art, media, science, technology . . .

10. Economic development lite: communication, art and ICTs in a globalised economy

Roger Sugden, Robbin Te Velde and James R. Wilson

1. INTRODUCTION

In a world characterised by 'globalisation', communication processes are a crucial determinant of the potential for economic development, as well as of the realisation of that potential. This is seen most straightforwardly in the influence of new technologies.

Although globalisation is a highly contentious concept, partly due to its being a multifaceted phenomenon that makes it difficult to analyse within one discipline, it clearly relates to the emergence of relationships that transcend previous boundaries (Massey, 1991; Radice, 2000; Scholte, 2000; Waters, 2001). In turn, the importance of transcending boundaries is associated with advances in information and communication technologies (ICTs), which have markedly eased previous difficulties inherent in interaction over large distances. The evolution of email and the Internet, for example, herald the cheap and almost instantaneous transfer of vast amounts of information across the entire accessed world. This creates the potential, increasingly being realised, for a new 'layer' of market and non-market activities that are detached from physical localities.

However, ICTs are not the only way in which communication processes impact on economic development, and in this chapter our focus also includes analysis of art in Britain in the 1990s, alongside evidence about the influence of large transnational corporations in schools and colleges, and in the film, television and magazine sectors. The particular aim of the chapter is to identify specific issues around communication that are at the heart of understanding the prospects for economic development. Part of its novelty is that the chapter covers very wide ground – from models of economic development to analysis of 'yBas' ('young British artists') – but its focus is actually on two possibilities. On the one hand, of economic development based on 'democratic globalisation' and, on the other hand, of development based upon exclusive governance (Sugden and Wilson, 2005). We hypothesise that these possibilities correspond to quite distinct approaches to communication.

The analysis is organised as follows. To provide a foundation, section II presents a particular perspective on economic development through a series of illustrations. section III considers the relationships between different models of economic development and different communication processes, explored in particular through an analysis of art in Britain in the 1990s. Section IV identifies various obstacles to achieving inclusive development, giving illustrations from popular discussion of globalisation. Section V returns us to a discussion of ICTs. Here it is argued that while the Internet might be an exceptional case of a true 'technology of freedom', there remain choices in how it is used if its full potential in stimulating and aiding democratic forms of globalisation is to be realised. Finally, section VI provides a summary of the discussion.

2. ECONOMIC DEVELOPMENT UNDER GLOBALISATION: GOVERNANCE, KNOWLEDGE AND LEARNING

Think for a moment about a question: 'What does the term "economic development" mean to you or to someone else who lives in the same locality as you"? Perhaps it means a desire that your children enjoy better education and/or healthcare. It might mean the opportunity for you to earn a higher income than at present. If you have a job, maybe you desire the creation of more jobs, or of better quality jobs, so as to alleviate poverty and social problems around you? 'Economic development' might also incorporate a wish to build on the traditions and cultures of your community. Perhaps you have an attachment to certain activities and ways of doing things that reflect the industrial history of where you live?

While we could (make an informed) 'guess' at the meaning of economic development for you, as 'outsiders' to your specific context we could not hope to be completely accurate. Within your locality itself there will be a range of different views, but there are also likely to be common concerns and desires. Indeed, people are inherently related through their direct and immediate stake in the development of the place or places where they live and/or work; they are linked through their everyday relationships and communications and through the common firms and institutions that impact on their daily lives. Moreover, many people in a particular place will have lived through its recent history and been brought up within its cultures and traditions. Consequently, there are likely to be common threads to development aims in particular places, and a potential for the emergence of broad consensus on what is desired, albeit one that is evolving and at times contested. We suggest that the conceptualisation of 'development' itself is necessarily contextual, and should start from these very aims and objectives. Following 'external' definitions and evaluations of development is problematic as there is a danger that the process of development comes to lack meaning for the people living and/or working in a particular locality (Sugden and Wilson, 2002). The idea is that external criteria inevitably reflect external interests, to that extent yielding inefficient outcomes with respect to the aims of the communities seeking to develop.

Now consider for a moment what 'economic development' means for those who make the key decisions impacting your locality. While some decisions are taken at local level, there are many decisions that are made by national governments or by firms headquartered outside your locality. Moreover, these external decisions are often 'strategic' in the sense that they determine overall direction, essentially setting the parameters in which local decisions can be made. While these decisions are likely to have significant impact on 'economic development' in your locality, the motivations and desired outcomes behind them may differ from those found locally. Indeed, what constitutes 'economic development' for the interests controlling firms or national governments, for example, are likely at times to cut across the objectives of those within particular localities. Furthermore, a combination of physical/cultural/mental (and power) distance between people in different places may render the co-ordination, transparency and accountability in such decisions relatively weak.

Many key choices with regards economic development strategy are also taken within international institutions such as the IMF and World Bank. The impact of strategy formed in Washington, for example, has been felt in localities around the world; directly, though externally directed programmes; and indirectly, through the concept of aid conditionality applied to national governments. Again, there are likely to be differences between the outcomes that constitute 'development' for those designing policy strategies in Washington, and the outcomes that are meaningful for people in diverse localities around the world. Once more, there are few, if any, channels for the people of those localities to have an influence.

Such institutions have been associated most especially with perspectives on development (and associated measures of economic progress, competitiveness, etc.) that are centred essentially on income (Wilson, 2008). Seers (1972, pp. 21–22) questioned this restriction: given that 'the complexity of development problems is becoming increasingly obvious, this continued addiction to the use of a single aggregative indicator . . . begins to look like a preference for avoiding the real problems of development'. Yet several decades later this restricted approach is being maintained, albeit increasingly questioned (Wilson, 2008; see also Rodrik (2008) on the prospects for a new Washington consensus based around each economy's particular circumstances, but with little mention of the fundamental *aims* of development as such).

The significance of decision-making processes has also been identified in more specific institutional contexts relevant to the development process. Consider, for example, the governance of firms and in particular transnational corporations (Cowling and Sugden 1998, 1999; Branston et al., 2006). It has been argued in this context that concentration of strategic decision-making power in the hands of a few implies a failure to govern production in the interests of the community at large (Cowling and Sugden, 1999). To avoid such failure it has been suggested that the democratisation of governance is required, to reflect the concerns of everyone interested in a corporation's activities and, more specifically, to enable governance in the interests of publics (Branston et al., 2006).

Accordingly, the implication of this perspective is a need for people to be involved in the planning for their development, and thus in the decision-making processes affecting the societies where they live. More generally:

> The prospects for economic development depend crucially on the governance of economic development.

Inextricably linked to the centrality of governance is a concern with knowledge and learning, the foundations on which engagement in governance can be built. However, here we would make a distinction between *knowledge for governance* and more traditionally analysed forms of knowledge. Compare, for example, the analysis of mainstream development economics. Given movement towards a 'new' economy, where knowledge is seen to be the key to realising returns, it has become widely accepted that education and the generation and diffusion of knowledge are vital components of development processes. As far back as the late 1990s, the *World Development Report* (World Bank, 1998), for example, was dedicated to knowledge. The standard approach has been to see knowledge as important in terms of the value it can add to production processes, and crucially, therefore, the ability to attract investment. A primary concern has been the bridging of knowledge gaps so that localities can 'compete' on a more equal footing, or the extending of knowledge gaps so that localities can enhance their 'competitive advantage'.

While we recognise the importance of this 'competitive' imperative for knowledge, our emphasis is different. We suggest that knowledge and learning are especially important when they take forms that afford people the opportunity to engage effectively in the decision-making processes that govern development. Appropriate knowledge and learning processes can free people to become suitably involved in formal and informal networks for the governance of firms, institutions, government and other economic actors. This is in part knowledge about how to govern democratically; a process of learning about how to form opinions and express them, and in doing so to impact on development. Where effective governance networks do not exist, knowledge and learning about governance, obtained through participation in and experience of the process itself, can lead to their establishment. Indeed, the economic benefits from such processes of active citizenship are not reliant on specific formal skills. They can, however, benefit from different types of knowledge in a whole range of fields, and certain forms of static information are important; data about decisions being made by firms or government are necessary if people are to develop views on and govern what happens in their societies. But most vital for success in this perspective is a dynamic process of learning to develop and communicate ideas and, in so doing, learning to determine democratically the development of a locality.

While learning to be democratic may appear a simple notion, it is an emphasis that is typically lost in debate and policy around knowledge and learning. However, such a focus is central to the approach to globalisation elaborated in Sugden and Wilson (2003, 2005). Here, knowledge and learning are seen to be especially pertinent in today's world because they determine how people are able to respond to and shape changing relationships and territorial boundaries. In this setting 'who' governs and 'how' they govern implies that globalisation can take two broad forms, 'elite' or 'democratic'. One is a process harnessed to further the prosperity of exclusive interests, the other a process harnessed to further the prosperity of all. Accordingly:

> If knowledge and learning are key aspects of the governance process, they are fundamental to the choice between elite and democratic globalisation.

3. ARTISTIC PROCESSES, COMMUNICATION AND ECONOMIC DEMOCRACY

It is in the context of the governance of economic development and the associated different possibilities for globalisation processes that we seek to analyse the role of communication. The attainment of democratic globalisation requires that each interested person be able to think about and participate fully in the governance of economic development. Each must be active, alert to the necessity for certain types of knowledge and learning; able to use their voice in the pursuit of that knowledge and learning; able to use that knowledge and learning in influencing strategic ways forward. This implies a need for discussion and interaction, for a conveying of information, ideas and feelings. Therefore appropriate and effective communication is necessary; because the art of communication is central to the art of democracy,[1] it is central to an understanding of the potential for economic development, as well as to the realisation of successful development.

Information, ideas and feelings can be communicated through various means: language, speech, prose, film, poetry, music, visual art and so on. It would seem likely that democracy in economic development requires a certain kind of freedom and opportunity for all people to express themselves as they wish. We hypothesise that freedom and ability to participate fully in economic governance and in the corresponding democratic knowledge and learning processes would reflect and be reflected by a parallel freedom and ability in communication media.

The possibility of a relationship between economic and communication processes is suggested by comments on Dutch artist Johannes Vermeer. It might be argued as no coincidence that in the seventeenth century Vermeer had an opportunity to express himself in his art in innovative ways at a time when the Netherlands was able to develop its economy in new directions.[2] Comments on, for example, film releases and on art nouveau also illustrate the possibility.[3] We will focus our illustration, however, on art in Britain in the 1990s.

Without suggesting that British art in that period was 'better' or 'worse' than at other times – that is not our concern at all – we would hypothesise that it did not provide the sort of art that is required by the type of democratic economy that we advocate. This is also not to comment on whether British art in the 1990s was in some sense closer to, or more distant from, what is required by a democratic economy, compared to other forms of art at other times.

In the 1990s 'Britain, or more accurately London, came to be regarded as the world capital of visual arts' (Spalding, 2003, p. 86). The focus of attention was a phenomenon that began to emerge in the 1980s and that has been variously labelled 'the "new British art", or simply the "new art", "Brit art", or even the "New Boomers", though the term that has stuck is "young British artists" (often abbreviated to "yBas")' (Stallabrass, 2006, p. 2). The label that Stallabrass himself prefers is 'high art lite' (ibid.), in part as that reflects the fact that the artists prominently associated with the phenomenon lived in Britain but were not exclusively British, and that such art was actually not confined to Britain (see also Spalding (2003), positioning yBas in the context of modern art more generally).[4]

According to Millard (2001), yBas gave especial stress to the marketing and branding of art. She argues that its stars such as Damien Hirst and Tracey Emin deliberately became brands, and she reports the views of Christopher Frayling, Rector of the Royal College of Art. He observes that the 1980s witnessed a seed change in approach characterised by an emphasis on marketing and 'a convergence of advertising and art' (p. 53). Within this, art students were taught business skills and these 'children of Thatcher' (p. 53) learnt how to market their products:

> I remembered artists in the 1970s who thought their role was to put two fingers up at the establishment and deliberately produce art which was unexhibitable. They didn't want to be marketed. But these [post 1970s artists] were the children of Thatcher. There was an idea that they ought to be more hip to the real world. I was swamped. They wanted to know how to get on, how to charge commission, how to get an exhibition. (p 53)[5]

Given that the then British Governments were stressing an explicitly free market approach to the economy, an implication might be that a market-based economy corresponded to a market-based art.

Whilst a casual and naïve view might be that artists only work from and for the soul, reality might be different; in market-based economies it appears that they might work from and for the market. Accordingly we would ask: if there is elite globalisation, do artists serve the latter and undermine democracy? Can democratic globalisation exist alongside artists obsessed with marketing and branding? We would argue that a healthy economic democracy requires art that is questioning and stimulating, but if artists become too close to business and business techniques they will have no independence and therefore damage the prospects for democracy. (This is not to imply that an artist must or should exist to serve democracy, or indeed to serve anything outside him or her self. Rather, it is to hypothesise a consequence of particular circumstances.)

The perverse and pervading ways that art might use business and business might use art are explored in Stallabrass (1999, 2006) and detailed (for the US as well as Britain) in Wu (2002) and Spalding (2003). Stallabrass (1999) argues:

> There was a strong radical strand in British art of the 1970s, producing work of an explicitly political kind (and such work was always, naturally enough, more dependent upon state patronage than the market); in the 1980s this persisted in some highly theoretical forms, though it was pushed to the sidelines by a buoyant private market that favoured bombastic neo-expressionist work; in

> the 1990s high art lite has produced a knowing synthesis of the two, combining idea and expression, theoretical niceties and visual spectacle in a neat, market-friendly and thoroughly apolitical package. (p. 173)

He identifies a coming together of state and private involvement, governments having brought the market into the public sector through, for example, corporate sponsorship of public exhibitions. Wu (2002) concludes that by the close of the 1980s art museums (in the US and Britain) 'had become just another public-relations outpost for corporations' (p. 122), and for Spalding (2003):

> The only way that the public's interest in art can be properly protected and served is by curators actively and disinterestedly seeking the best art they can find. Today's curators, however, have ceased to be the public's representatives at the court of art; instead they have become at best artists' agents, at worse dealers' foot soldiers. (p. 90)

According to Stallabrass (1999), corporate sponsorship has affected both the content and presentation of art:

> It makes a difference to the art displayed to brand an exhibition with the logo of a main sponsor, just as it does to sell coasters or carrier bags bearing reproductions of the exhibits. The more exhibition-going has come to be seen as an extension of shopping, a leisure activity rather than an educational one, the more it can feature in the lifestyle magazines. While such developments well predate high art lite, the tendency is highly suited to this new environment and makes the relation between art and commerce more visible, being a postmodern paradigm of such consumerism, especially in its unconcern with the activities of business. (p. 173)

Art as shopping rather than communication[6] is in line with a free market economic process, but it is not what is required for economic development based on the democratic pursuit of democratically identified aims and objectives.

The deficiency is suggested by the ways in which elites may control such market-based art. In a sense, yBas was inclusive and apparently anti-elitist; it purported to bring high art to the masses. However, it has been argued that the masses were included on terms that were set by exclusive controlling (governing) interests. Spalding (2003) refers to the especial influence of 'a tiny handful of people in powerful positions' (p. 86), and indeed goes so far as to highlight the especial power of just one man: Charles Saatchi, who 'exerted great influence from the 1980s onwards simply by becoming by far the biggest buyer, exhibitor and promoter of modern art' (p. 86). This parallels the way in which free market economies might be argued to be a force for inclusion across the world, but a force that includes on restrictive terms, and most particularly terms that deny inclusion in strategic decisions, for example, decisions about the aims and objectives of economic activity. In turn, this is similar to the way in which transnational corporations may be seen as a force for industrial development, but one that is on the terms of a strategic decision-making elite concentrated in a handful of the world's major cities (Hymer, 1972).

Stallabrass (1999, p. 194) sees a contradiction within the British Government in the 1990s, a 'dangerous game, fostering political reaction' because it pursued policies of 'cultural liberalism' that promoted inclusiveness alongside policies of 'economic neoliberalism' that implied widening economic inequality. What Stallabrass misses is that whilst those economic policies might have resulted in inequality, they were nonetheless presented using rhetoric of inclusion. That was arguably no different to what Stallabrass observes of reality in the art world; government policies apparently designed to make high art inclusive but in fact driving out high art, bringing psuedo-inclusion into something different and actually implying a second rate form of communication. High art lite accompanied by 'economic development lite'?

For Stallabrass (1999), the art market 'is highly controlled, not just in terms of production, as one would expect, but also in terms of consumption' (p. 182). He tells of private galleries maintaining a close-knit pool of buyers and being uninterested in selling outside that pool: 'it is important for a gallery to know that they are selling to someone who is serious about collecting – that is, someone who is a consistent buyer of work in a certain price range and who can be relied upon not to dispose of the work irresponsibly' (p 182). Galleries also buy work from 'their' artists at public auctions, thereby restricting supply. And as for production, Stallabrass argues that galleries have significant influence over 'the type of work, the amount of it, the size of editions and the setting of prices' (p 182). Whilst this is not a new characteristic for the 1990s, he points to special difficulties. One requirement is that galleries like their artists to continue to produce the same kind of work, seen to be less of a problem for previous generations because the nature of their art required decades to perfect their skills:

> The idea is that if you keep plugging away at a single trick for long enough, the buyers (some of whom are a little slow) will get it, and in any case it becomes recognisable and therefore accessible through its very familiarity. By insistent repetition, the artists in effect brand their work. (Stallabrass, 1999, p. 182)

We hypothesise that elites seeking to control art markets in these ways do so for their own self-interest (in line with the standard economist's assumption);

that would be at odds with art as a catalyst for democratic thought and expression. Stallabrass (2006) makes a similar point more colourfully:

> An essential question – and it is an old one, present at the inception of democracy – that should be asked of an art that deals with mass and popular culture, that hails the non-elite audience, is whether it takes as its task to stir 'hearts and to prevent them falling asleep in that false and wholly material happiness which is given by monarchies'.[7] The majority of artists purveying high art lite, shielding their conservatism with the appearance of the modern and the radical, have been content to play the well-remunerated role of court dwarf. (p. 308)

4. TRANSNATIONALISM AND 'FREE' MARKETS

Further obstacles to achieving inclusive development and democratic globalisation are revealed by other ways in which the pursuit of markets and market practices, alongside existing islands of power, may impact on communication and the communication industries. Again some of the difficulties can be illustrated from the globalised art industry, although the problems extend much further.

Central actors in today's globalisation processes in practice are the transnational corporations, firms that each produce in and span various countries. When such firms dominate a communication industry, the result can be development in exclusive rather public interests. This seems to be a problem in the art world. Wu (2002) comments on the ways in which US museums have behaved as transnational corporations, apparently exploiting market power and constraining effective communication. She focuses on the Boston Museum of Fine Arts and in particular on the Guggenheim Museum, essentially seeing this as 'dominated by the ethos and practices' (p. 284) of transnational corporations.

On the activities of the Guggenheim, for instance, Wu reports:

> In addition to its Frank Lloyd Wright building in Uptown Manhattan, the Guggenheim owns a Guggenheim Museum SoHo branch in Downtown New York, the Peggy Guggenheim Collection in Venice, the Guggenheim Museum Bilbao in Spain and its latest edition, Deutsche Guggenheim Berlin in Germany. This list does not, of course, include other failed franchising attempts during the last decade. (p. 283)

She questions the ways in which it raised capital, its relationships with other organisations and the role it played in 're-branding the Basque capital' (p. 287). Wu concludes that 'globalism may have its validity as a utopian ideal, but the overseas expansion of American art institutions that we have witnessed over the last few years is the antithesis of idealism;

it has been an exercise in pragmatism in which cultural imperialism and multinational capitalism have joined forces to consolidate existing hegemonies' (p. 292). Such hegemony is not consistent with economic development based upon the democratic governance of the people of a locality. It implies communication that is dominated or controlled by corporations from outside. Regarding the Japanese branch of the Boston Museum of Fine Arts, for example, Wu argues: 'it is the American institution that has the final say on what art its Japanese clients will actually see, even if its curatorial decisions should prove to run counter to the wishes of the Japanese themselves' (p. 291). Similarly, she sees the Guggenheim as having placed control amongst exclusive interests in the US.[8]

More generally, there are wider queries over communication and communication industries controlled by elites, with knock-on effects to what is taught in schools, sold in shops, therefore to the sorts of discussion and debate most people are (not) opened up to. Branston and Wilson explore these issues in the context of the media industry in their contribution to this volume. Here we illustrate from reports in Klein (2000).[9]

She describes a form of what we categorise as elite globalisation. Her particular focus is the use of brands and branding by corporations that, as a consequence, allegedly exercise undesirable influences on people's lives. She considers their impact on schools and universities, including their influence on communication and the stimulation of communication in these contexts. This is doubly important for us because of the role of knowledge and learning in development and globalisation processes.

There is argued to be an impact resulting from sponsorship of various kinds. Klein contends that, 'regardless of the intentions when the deals are inked, the fact is that campus expression is often stifled when it conflicts with the interests of a corporate sponsor' (p. 97). Examples are provided of a speaker denied funding at Kent State University (US) because of a plan to talk negatively about Coca-Cola's activities in Nigeria, Coca-Cola having exclusive vending rights at Kent; and of an anti-smoking group denied permission to protest at an event sponsored by Imperial Tobacco at York University (Canada). She also points to more subtle but nonetheless significant effects. Echoing the suggestion that art has become a shopping rather than an educational experience, Klein (p. 98) refers to the 'slow encroachment of the mall mentality' in universities where large corporations spread their brands into each nook and cranny. In addition, when corporations fund research, they are said systematically to taint the message and sometimes to influence the output. Evidence of the latter is given from three cases involving medical researchers able to publicly challenge their sponsors and employers, the inference being that other instances are swept under the carpet.

Klein similarly alleges inappropriate use of market power by media and other corporations outside of education. It is argued, for example, that the desire to nurture a suitable brand stifles communication because it leads to incompatible material being excluded. This is said to have affected not merely what is available in shops but also what is produced to begin with. She expands by using examples of material withdrawn by retailers because it violates their 'family image'; Wal-Mart, Kmart and supermarket chains have rejected magazines and CDs because their content and covers were deemed unacceptable. Similarly, Blockbuster Video has rejected films rated unsuitable for under-17-year-olds. Klein also reports that many magazines have shown leading retailers advance copies of new issues before they were finalised for shipment, and that films have been produced in different versions, one for cinemas and another for Blockbuster. Likewise, CDs produced in two forms by established bands, leaving new entrants with the choice of no production versus sterilised production.

Concentration of power in the hands of retailers is not argued to be the sole issue. In practice retailers, distributors and producers are often governed by the same group. Common ownership of Paramount Films and Blockbuster Video, and of Disney, Miramax and ABC, are cited as examples. Also in the news and current affairs sector, where the potential consequences for communication are undoubtedly significant:

> As multinational conglomerates build up their self-enclosed, self-promoting worlds, they create new and varied possibilities for conflict of interest and censorship. Such pressures range from pushing the magazine arm of the conglomerate to give a favourable review to a movie or sitcom produced by another arm of the conglomerate, to pushing an editor not to run a critical story that could hurt a merger in the works, to newspapers being asked to tiptoe around judicial and regulatory bodies that award television licenses and review anti-trust complaints. (p. 169)

Klein's examples are drawn from ABC dropping a story concerning Disney, from cases involving the then Time Warner/Turner group, and from Rupert Murdoch cutting news items on China. The point is not that such activity is new, rather that the concentration of activities in the conglomerate implies problems of a different dimension.

The fundamental issue is that Klein presents evidence consistent with exclusive governance in the economies of the world, and inconsistent with inclusive, democratic economies.[10] In part this is because she is illustrating specific cases where exclusive interests appear to dominate. But most importantly, in the cases we have reiterated, it is because she is suggesting that *communication is distorted* to favour those exclusive interests, a conclusion that is consistent with comments in Lee et al. (2001) on news media. Communication is not serving the needs of democracy, despite its often being packaged in a language of freedom and choice.

A familiar response to this interpretation is that corporations only provide what markets demand; it is not the exclusive interests of corporations that are overriding, but the demands of consumers. The argument might be that although Wal-Mart has removed certain items from its shelves, it could only do so because that is what its customers wanted and demanded. However, a flaw in this response is that consumers are not sovereign in an economy where large (transnational) corporations have market power that can be wielded to manipulate and dominate their demands. Moreover, taking a wider viewpoint we would argue that socio-economic development and the well-being of people needs to be understood from the perspective of the sovereignty of people not merely as consumers but rather in all of their dimensions. That is what underlies the approach that we are advocating. Even if Wal-Mart had responded to and respected consumer demand, that would not have sufficed for good governance of the economy.

Sunstein (2001) analyses related concerns from the perspective of constitutional law, political democracy and communication in the new economy. He contrasts consumer and political sovereignty, arguing that an appropriate political democracy requires something other than the unfettered pursuit of consumer interests. In particular and unlike consumer sovereignty, political sovereignty 'does not take individual tastes as fixed or given' (p. 45). Following Branston et al. (2006), we would emphasise the same requirement for an economic democracy (and see also Adaman and Devine (2001) on deliberative democratic processes in economic planning). The implication, to parallel Sunstein, is that good economic governance (of firms, corporations, localities, nations and so on) needs discussion and reason-giving in the public domain.

There is no effective democracy if a decision is 'not backed by justifications, and represents instead the product of force or simple majority will' (Sunstein, 2001, p. 45).[11] Fundamentally, there is a need for 'institutions designed to ensure a measure of reflection and debate – not immediate responses to whatever people at any particular moment in time, happen to say that they want' (p. 197). Put differently, there is a need for a certain sort of freedom, one which 'requires exposure to a diverse set of topics and opinions' (p. 200). For Sunstein:

> This is not a suggestion that people should be forced to read and view materials that they abhor. But it is a claim that a democratic polity, acting through democratic organs, tries to promote freedom, not simply by respecting consumer sovereignty, but by creating a system of communication that promotes exposure to a wide range of issues and views. (p. 200)

We would argue that democratic globalisation similarly requires communication based on widespread exposure to diverse possibilities, a stimulating and catalytic communications industry that evidence appears to suggest is absent in today's free market economies.

A specific aspect of this requirement is the corresponding need for public space where economic citizens can interact and exchange ideas. Sunstein recognises this in the political sphere. He comments on the historical and continuing importance of 'public spaces' in a political democracy; in the US, for example, streets and parks have to be kept open for the public to be able to express their views, to assemble and communicate their thoughts and ideas. He also argues that this doctrine should be expanded, not least to the mass media and the Internet, where he sees promising opportunities but also risks. We would make a similar case for the economic sphere, although before becoming preoccupied with expansion it would be worth recognising misgivings about the way space can be taken over by the exclusive interests of corporations. Klein (2000) calls this the 'privatisation of public space' (p. 156). She discusses the development of 'branded villages' (p. 182), notably the Disney initiative in Celebration, Florida, but also interesting is the experience of malls in the US. For Klein, in becoming 'the modern town square' malls have 'created a vast grey area of pseudo-public private space' (p. 183). She argues that whereas traditional town squares 'were and still are sites for community discussion', speech in the malls is constrained; 'peaceful protestors are routinely thrown out by mall security guards for interfering with shopping, and even picket lines are illegal inside these enclosures' (p. 183).

We submit that this is not merely a political issue. It is also problematic for the requirements of inclusive economic development; democratic globalisation needs public space in which participants in the economic process can communicate with each other in appropriate freedom.

5. NEW COMMUNICATION TECHNOLOGIES

Our concerns regarding the link between communication, democracy and economic development are clearly influenced by technological developments, and especially by the phenomenal rise of the use of the Internet during recent years. In an orthodox (mainstream economics) sense, new ICTs might impact on economic development by opening the way to potentially greater efficiency, and therefore growth, among firms and institutions that are able to adopt them. However, they also have the potential to impact on economic development in a different way, through their influence on communication, and therefore governance. Alas, much of the gist of the complex interaction between society and technology seems to be missed because there remains a great divide between social scientists and engineers. The debate about the – presumed – influence of the Internet on patterns of communication is no exception to this rule. In this final section we make a modest attempt to at least partly bridge the gap between social theory and engineering practice.

The establishment of shared positive expectations about an innovation-to-be is part and parcel of the implementation of any new technology (Te Velde, 2004). While the realisation of a new technology requires considerable resources, a crucial challenge for the entrepreneur is to acquire these resources while there is yet little to offer in return to the investors (ibid.). In the case of new communication technologies, it was argued early in their evolution that there was an 'inherent anarchism' in these new networks conducive to freedom (De la Sola Pool, 1983). In general, one should be wary about such optimistic claims. Following the line of Sunstein (2001) and others, new technologies can be used either to serve the purpose of deliberative democracy or to serve populism, a position based on the basic assumption that technology is neutral ('a knife is a knife is a knife'). Thus it is the social context which ultimately determines the social impact of a technology.[12]

This, we would argue, is a rather naïve view. Technology is indeed always introduced against the backdrop of existing societal context, but it also has a pre-structuring effect in its own right (Akrich, 1992). Even a multi-purpose tool, such as a knife, facilities certain types of behaviour and hampers others. Although technological and societal progresses each have their own dynamics, there exists no dichotomy: technology *is* society, albeit of a different kind. It is a specific set of social practices ('ways of doing') frozen in time and embodied in a technological system (Te Velde, 2008). When a new technology is introduced in an existing societal setting it is really the principles of its designers which clash – or most likely fit – with the dominant principles in the particular societal setting (ibid.).

> Our claim here is that the design principles of the Internet are indeed 'inherently anarchistic'.

Thus in the particular case of the Internet, the optimistic claims of De la Sola might be justified after all.

The contrast between the design principles of the Internet (packet-switched) and of traditional (circuit-switched) telecommunication technologies clearly illustrates the point. In the latter case, there is strong congruence between the design principles and centralized decision-making structures. Both require high degrees of formalization and standardization, and objects that are difficult to 'informatize' (such as local practices) are regarded as nuisance or even threats to the process of modernization (Galjaard, 1979; Zuurmond, 1998). This is translated into configurations of organization and technical design where the intelligence is put in the core of the system and control in the hands of the 'experts'. In this sense the structure and functioning of traditional telecommunications systems closely resembles those of corporations. Both are large, single, hierarchical structures with a central brain and an ever more complex underlying system (Te Velde, 2008).[13] Although one could rightfully consider the current age of transmission as a radical break with the past in terms of magnitude of coordination and control (see for instance Castels, 1996), the principles and structures have hardly changed. Indeed, contrary to the argument of Wyner and Malone (1996), the sharp decline in communication costs do not imply a shift from centralised to decentralised decision-making structures, but rather a rebirth of centralised decision-making in a technologically superior form (Te Velde, 2008).

The technical regime of the Internet is a clear exception to the centralised 'way of doing'. The founding fathers of the Internet wilfully designed the technology so that the intelligence is at the fringes of the system, and thus control in the hands of the end-users. While mainstream engineers and scientists initially ridiculed the system, its technical characteristics enabled the establishment of a decentralised, anonymous virtual communications network that can operate over any existing telecommunications system and in a highly cost-efficient way. Hence, with relatively little investment, end-users can communicate with each other from virtually any underlying physical network, apparently without being traced, censored or disturbed by a centralising controlling agency (Sugden et al., 2002).

The contrast between the 'decentralised' user-driven regime of the Internet and the 'centralised' corporate-driven regime of traditional telecommunications is vividly reflected in the current fierce debates about 'net neutrality'. What appears to be a purely technical issue is essentially a fundamental ideological clash. The net neutrality principle states that '[a] neutral broadband network is one that is free of restrictions on the kinds of equipment that may be attached, on the modes of communication allowed, which does not restrict content, sites or platforms, and where communication is not unreasonably degraded by other communication streams' (Wu, 2008). Opponents of the principle have argued that the Internet has become so congested that the only way to guarantee reasonable quality of service is to prioritise certain types of data traffic, that is, to discriminate between different types of data traffic. However the only way to do so is to monitor traffic. Until fairly recently, traffic monitoring could be limited to the inspection of the headers of data packets. However the types of data traffic that corporations would like to discriminate against (esp. 'peer to peer' or 'P2P') tend to disguise themselves in all kinds of creative manners. Therefore simple header inspection is no longer sufficient, and the latest generation of traffic monitoring ('deep packet inspection') is able to look into the data in real-time (Brennenraedts at al., 2008).[14] The use of deep packet inspection is mostly justified by purely technical motives ('optimisation of traffic'), but obviously also opens up all kinds of interesting political, commercial and control uses for a centralised decision-making structure.

The introduction of the Internet in China is an interesting case in this respect. The truly decentralised nature of the Internet makes it almost impossible to control.[15] Nevertheless the Chinese government is taking great lengths to protect its citizens from the alleged evils of pornography, Falun Gong or democracy, for example.[16] All chat rooms and discussion boards are being moderated, any visitor to a cyber café has to register beforehand, and websites of a sensitive nature are being blocked.[17] Foreign-owned Internet companies such as Yahoo[18] and Google[19] also comply to the official Chinese standards.

It is technically impossible to control all data traffic going in and out of China, even with advanced methods such as deep packet inspection. The amount of data involved is simply too much for any 'Great Firewall of China' to deal with. Furthermore, there are also technical countermeasures to intelligence operations on the Internet. Virtual tunnelling, for instance, keeps private data safe from deep packet inspection.[20] Nevertheless, it remains to be seen to what extent such advanced methods can be applied by average citizens. Indeed, the control of the government does not have to be foolproof. As long as the public at large is not sure what is technically possible and what is not, and there are occasional Internet-based arrests, such attempts at censorship of communication are likely to have an important impact.

Despite these controlling tendencies, it is undoubtedly the case that the Internet enables decentralised, anonymous global communications at virtually no cost, as a consequence turning every citizen into a potential reporter. Indeed, the viral character of information on the Internet – as witnessed in the case of the Chinese gymnast – is a major nuisance to centralised decision-making structures.[21] The impressive army of 30000 censors could be offset if 0.002 per cent of the Chinese population turns into self-appointed news agents, for example. Moreover, the increasingly *collaborative* efforts of all Internet *pro*sumers – the much-hailed Web 2.0 revolution, including for example the development of social networking sites, blogs, wikis and the like – holds even greater potential for processes of democratisation.

Having said this, the technical possibilities of the Internet do not automatically give rise to more appropriate and effective communication, and thus enhanced participation in the governance of economic development. Authors such as Keen (2007) have even argued that such 'democratisation' of the media is a change for the worse. The vast army of amateur journalists, artists, movie directors and programmers is creating 'an endless digital forest of mediocrity' (Keen, 2007, p. 3). According to Keen, then, the 'wisdom of the crowd' is rather shallow or even downright dangerous, since the rise of the amateurs erodes the position of the skilled professionals.

The conservative Keen may have a point in stating that the culture of the masses might not be the 'lofty culture' of intellectuals.[22] Yet there is no reason why amateurs and citizens should be less able to communicate than professionals and politicians. As Lawrence Lessig has forcefully argued, the traditional system can be just as bad as the worst of the Internet (Lessig, 2007). Ultimately, what determines the quality of knowledge creation is the extent to which the process of knowledge creation is truly open and inclusive, not so much the qualifications of the people involved. The major advantage of the Web 2.0 revolution is exactly the massive numbers of people that are involved in supposedly collaborative efforts. Obviously, sheer numbers alone will not do the trick – the open source adage that given enough eyeballs, all bugs are shallow does not hold per se. Yet with many eyes, shallow bugs get caught very quickly, and the more eyes there are, the more likely it is that *some* member of the group has sufficiently penetrating vision to catch the deeper-swimming bugs (Chandler, 2008).

To sum up, given the peculiar way in which the Internet has been designed, it might be one of the exceptional cases of a true 'technology of freedom'. Technology alone will of course not lead to more liberation and open debate, but the principles do at least have a bias against centralised decision-making structures. In this respect, social scientists and citizens alike should take due care to the ongoing net neutrality debate and not leave this fundamental issue to engineers alone.

6. CONCLUSION/SUMMARY

We see communication as influencing economic development, and vice versa. Because the art of communication is central to the art of democracy, it is fundamental to development. In particular, it would seem likely that democracy in economic development requires a certain kind of opportunity for all people to express themselves as they wish. Freedom and the ability to participate fully in economic governance, and in the corresponding democratic knowledge and learning processes, would hence reflect and be reflected by a parallel freedom and ability in communication.

The possibility of a relationship between economic and communication processes is borne out by comments on art in Britain in the 1990s, and we conjecture that this period of so-called high art lite corresponded to an age of economic development lite. High art lite was apparently inclusive and anti-elitist, purporting to bring high art to the masses. Yet those masses were included on specific terms set by a controlling interest group. This parallels the way in which 'free' market economies might be argued to be a force for inclusion across the world, but a force that includes on restrictive terms, and most particularly terms that deny inclusion in strategic decisions, for example, decisions about the aims and objectives of economic activity.

It has been argued that Britain in the 1990s was characterised by art as shopping, not art as communication. Such a focus is in line with a free market economic process, but we query its implications if the intention is economic development based on the democratic pursuit of democratically identified aims and objectives.

More generally, there are further ways in which market practices may impact on communication and communication industries, undermining the possibility for achieving democratic globalisation. This can be illustrated by evidence on the influence of large transnational corporations in schools and colleges, and in the film, television and magazine sectors. We argue that economic development and the well-being of people need to be understood from the perspective of the sovereignty of people, not merely as consumers but rather in all of their dimensions. Good economic governance needs discussion and reason-giving in the public domain, a form and structure that enables and fosters deliberation and debate, a freedom to be exposed to diverse opinions and a wide range of issues. It needs a stimulating and catalytic communications industry of a type which evidence suggests is not necessarily a feature of free market economies.

As for ICTs, they have been welcomed as harbingers of a new democracy, but our conclusion is that there are challenges to be overcome if they are to be the servants of inclusive economic development and not mastered by exclusive interests. The underlying technology of the Internet has the potential for a decentralised and anonymous communications network, relying flexibly on a relatively simple infrastructure. However, there is not yet universality of access, and in many cases there are filter mechanisms that might militate against deliberation and open debate, including considerable opportunity for censorship and control. For the development of the art of communication in ways that strengthen the art of democracy, it is imperative that we take advantage of the 'bias' that Internet technology has presented against centralised forms of control. This means guarding against tendencies towards centralisation of Internet control among those groups (governments, large firms) that have an interest in controlling communication.

ACKNOWLEDGEMENTS

The authors would like to thank David S. Allen, J. Robert Branston, Silvia Sacchetti, Marcela Valania and Jutta Vincent for helpful advice and comments on earlier drafts and ideas. Earlier work was presented at the Conference on 'Transmissions: Technology, Media, Globalization', University of Wisconsin–Milwaukee, April 2002, and at the EUNIP International Conference, Turku, December 2002; we are grateful to participants for discussion.

NOTES

1. Dewey (1916, p. 7) argues that 'all communication is like art' and Hirschman (1970) discusses the art of democracy.
2. See Will Hutton's article on the future for Britain, 'Visions of Johannes', *The Observer*, 24 June 2001. He concludes: 'Britain has reached an important juncture. We are impatient for real success and have broken out of the ideological categories that have boxed in our thinking. Delft and seventeenth-century Holland weren't socialist, statist or inefficient – they allowed their civilisation to express itself publicly. And they gave us Vermeer. Unless we learn the same lessons, contemporary Britain will bequeath future generations nothing'.
3. On the influence of marketing and branding in the film industry, see 'Get set for the rehashes and spin-offs', *International Herald Tribune*, 20-21 April 2002, and 'Franchise fever', *Newsweek*, 22 April 2002. On art nouveau see Greenhalgh (2001): 'art nouveau was a style dependent on a market committed to the consumption of modern luxury goods' (p. 16).
4. 'Around 1999, a shift of attitude occurred in the British art world, and high art lite lost both its position at the cutting edge of the scene, and its immunity from criticism' (Stallabrass, 2006, p. 286).
5. This approach has not been confined to the Thatcher years. More recently, for example, the Slade School of Fine Art in London appointed a former investment banker to provide artists with business and management techniques (for a somewhat sceptical report see 'Art school turns to former banker to give their students an eye for figures', *The Independent*, 9 March 2002).
6. 'Art is essentially a means of visual communication, though that does not mean to say that all visual communications are works of art. To qualify for such an elevated status they must convey content of lasting value. . . . We reserve the word "art" for those rare visual creations that stir our emotions and stimulate our thoughts profoundly and elusively, which we find difficult to express through other means, but which we nevertheless feel to be true to our experiences. . . . Those accepting such a definition must be discontented with most of the art they see in public galleries today' (Spalding, 2003, p. 12).

7. Stendahl, 1839, *The Charterhouse of Parma*; p. 447 of the translation by C.K. Scott Moncrieff, 1931, London: Chatto and Windus.
8. There are parallels between the art world and academe, see Sugden (2004) on universities copying transnational corporations.
9. Klein's popular discussion might be thought to reflect at least aspects of so-called anti-globalisation campaigners. In using it to illustrate an absence of democracy, we are essentially pointing to our perspective on globalisation and to its foundations as the basis for a conceptual and theoretical analysis that explains and explores the concerns of many protestors on the streets of Seattle, Genoa, and wherever there is a meeting of the forces at the core of the so-called Washington or post-Washington consensus. Moreover, because these conceptual and theoretical foundations also provide the basis to explain and explore the views of the 'consensus', they constitute a framework that might enable the two sides to appreciate and understand the perspectives of each other.
10. Returning to our concern with art, Klein also discusses constraints on artists' use of corporate logos in their material; whilst logos might be seen as part of the language by which people communicate, and indeed that is arguably an aim for their creators, if artists use logos in material that the governors of the corporations find unacceptable, they are subject to constraint using trademark, copyright and libel law.
11. This is an argument against populism masquerading as democracy. Consumer sovereignty and good governance of the economy may be in tension if, for example, consumer sovereignty implies inadequate debate about, and understanding of, issues that are the subject of economic governance. This can also be seen in terms of freedom, again drawing on Sunstein: 'freedom consists not simply in preference satisfaction but also in the chance to have preferences and beliefs formed under decent conditions – in the ability to have preferences formed after exposure to a sufficient amount of information, and also to an appropriately diverse range of options' (p. 50).
12. The view that technology is value-neutral has been forwarded by pre-eminent thinkers such as Habermas (1970), and is still shared by many social scientists today.
13. The two systems also have a lot of overlap in geographical terms. The topology of the global communication systems mirrors the layered uneven power distribution of global capitalism as outlined in Hymer (1972). The hub-and-spoke structure of the system is based upon a core of interconnectivity between world cities on coastal shores, and these are exactly the same cities where the majority of major transnationals have located their headquarters.
14. The latest deep packet inspection software is able to monitor data at a very detailed level, such as a specific movie title within a stream of P2P traffic. See, for instance, Ipoque (2007).
15. As witnessed, for instance, by the embarrassing unveiling of an official document from the Chinese government which stated the true age (14) of one of its gold-winner gymnasts (the minimum age to compete for gymnasts is 16 years old). See http://strydehax.blogspot.com/2008/08/hack-olympics.html for a lucid report on the case.
16. It is estimated that some 30 000 Chinese civil servants are monitoring Internet traffic and blocking content that is deemed undesirable. Keywords such as 'democracy' or 'falun gong' in a search engine result in an error message (http://www.greatfirewallofchina.org).
17. To test real-time whether a website (e.g. freetibet.org) is being blocked by the Chinese government, use the tool at http://www.websitepulse.com/help/testtools.china-test.html.
18. According to Reporters Sans Frontières, in 2003 dissident Jiang Lijun was sentenced to four years imprisonment for 'undermining the state'. His conviction was based on a draft email found on his Yahoo page. This draft contained proposals for a more democratic China, which, according to the prosecution, could be regarded as taking part in 'subversive activities that aim to undermine the authority of the Communist Party'. Yahoo provided the necessary data to find Jiang (source: http://www.greatfirewallofchina.org).

19. References to the youthful Chinese gymnast (see note 15) in the cache from Google have also been erased.
20. See for instance the Tor project, www.torproject.org.
21. Amnesty International exploits this viral character in its 'irrepressible' initiative, see http://irrepressible.info/.
22. Recent field research on the use of webcasting by local councils suggest that dirt (Australia's MP Kevin Rudd's infamous ear wax footage – over 250000 on YouTube) and sex still sell. In the latter case, two local Dutch politicians were caught having sex on a security cam in a bicycle shed. The next webcasting of the local council meeting, which usually attracts around 100 viewers, this time attracted 25000 viewers. Eventually the server crashed because of overload (Brennenraedts et al., 2007).

REFERENCES

Adaman, Fikret and Devine, Pat (2001), 'Participatory planning as a deliberative democratic process: a response to Hodgson's critique', *Economy and Society*, **30**(2), 229–39.

Akrich, Madeline (1992), 'The de-scription of objects', in Wiebe E. Bijker and John Law (eds), *Shaping Technology/Building Society: Studies in Sociotechnical Change*, Cambridge, MA: MIT Press, pp. 205–24.

Arrow, Kenneth J. (1991), 'Scale returns in communication and elite control of organizations', *Journal of Law, Economics, and Organization*, **7**, special issue with papers from the Conference on the New Science of Organization, January 1991, 1–6.

Branston, J. Robert, Keith Cowling and Roger Sugden (2006), 'Corporate governance and the public interest', *International Review of Applied Economics*, **20** (2), 189–212.

Brennenraedts, Reg, Christiaan Holland, Ronald Batenburg, Pim Den Hertog, Robbin Te Velde, Slinger Jansen and Sjaak Brinkkemper (2008), *Go with the Dataflow! Analysing the Internet as a Datasource (IaD)*, The Hague: Ministry of Economic Affairs.

Castells, Manuel (1996), *The Information Age: Economy, Society, and Culture*, Oxford: Blackwell.

Chandler, Raymond (2008), 'Many eyes, shallow bugs – the untold story', accessed 2 September at http://lists.gnu.org/archive/html/emacs-devel/2008-01/msg00728.html.

Cowling, Keith and Roger Sugden (1998), 'The essence of the modern corporation: markets, strategic decision-making and the theory of the firm', *Manchester School*, **66**(1), 59–86.

Cowling, Keith and Roger Sugden (1999), 'The wealth of localities, regions and nations: developing multinational economies', *New Political Economy*, **4**(3), 361–78.

De La Sola Pool, Ithiel (1983), *Technologies of Freedom*, Cambridge, MA: Bellknap Harvard.

Dewey, John (1916), *Democracy and Education*, New York: Macmillan.

Docherty, Thomas (2008), *The English Question. Or Academic Freedoms*, Brighton: Sussex University Press.

Galjaard, Jacob (1979), 'Informatisering: paradox van organisatietechnologie', PhD thesis, Delft University of Technology, Delft.

Greenhalgh, Paul (2001), *The Essence of Art Nouveau*, New York: Harry N. Abrams.
Habermas, Jürgen (1970), 'Technology and science as "Ideology"', in *Toward a Rational Society; Student Protest, Science, and Politics*. Boston, MA: Beacon Press.
Hirschman, Albert O. (1970), *Exit, Voice, and Loyalty. Responses to Decline in Firms, Organizations, and States*, Cambridge, MA: Harvard University Press.
Hymer, Stephen H. (1972), 'The multinational corporation and the law of uneven development', in: J.N. Bhagwati (ed.), *Economics and World Order*, London: Macmillan.
Ipoque (2007), 'Internet study 2007', Leipzig: Ipoque. accessed Autumn 2008 at www.ipoque.com/news_&_events/internet_studies/internet_study_2007.
Keen, Andrew (2007), *The Cult of the Amateur: How Today's Internet is Killing Our Culture*, New York: Doubleday/Currency.
Klein, Naomi (2000), *No Logo: No Space, No Choice, No Jobs*, London: Flamingo.
Lee, Chin-Chuan, Zhongdang Pan, Joseph Man Chan and Clement Y.K. So (2001), 'Through the eyes of US media: banging the democracy drum in Hong Kong', *Journal of Communication*, pp. 345–65.
Lessig, Lawrence (2007), 'Keen's "The Cult of the Amateur": BRILLIANT!', accessed 2 September 2008 at http://lessig.org/blog/2007/05/keens_the_cult_of_the_amateur.html.
Massey, Doreen (1991), 'A global sense of place', *Marxism Today*, June, pp. 24–9.
Millard, Rosie (2001), *The Tastemakers. UK Art Now*, London; Thames and Hudson.
Radice, Hugo (2000), 'Responses to globalisation: a critique of progressive nationalism', *New Political Economy*, **5**(1), 5–19.
Rodrik, Dani (2008), 'Spence christens a New Washington consensus', *Economists' Voice*, **5**(3), Article 4, accessed at http://www.bepress.com/ev/vol5/iss3/art4.
Scholte, Jan Aart (2000), *Globalization: A Critical Introduction*, Basingstoke: Palgrave.
Seers, Dudley (1972), 'What are we trying to measure', *Journal of Development Studies*, **8**(3), 21–36.
Spalding, Julian (2003), *The Eclipse of Art. Tackling the Crisis in Art Today*, Munich: Prestel.
Stallabrass, Julian (1999), *High Art Lite. British Art in the 1990s*, London: Verso.
Stallabrass, Julian (2006), *High Art Lite. The Rise and Fall of Young British Art*, London: Verso.
Sugden, Roger (2004), 'A small firm approach to the internationalisation of universities: a multinational perspective', *Higher Education Quarterly*, **58**(2/3), 114–35.
Sugden, Roger and James R. Wilson (2002), 'Development in the shadow of the consensus: a strategic decision-making approach', *Contributions to Political Economy*, **21**, 111–34.
Sugden, Roger and James R. Wilson (2003), 'Economic "prosperity" and "globalisation": an agenda and perspective', in Roger Sugden, Rita Hartung Cheng and G. Richard Meadows (eds), *Urban and Regional Prosperity in a Globalised, New Economy*, Cheltenham, UK and Northampton, MA, USA: Edward Elgar.
Sugden, Roger and James R. Wilson (2005), 'Economic globalisation: dialectics, conceptualisation and choice', *Contributions to Political Economy*, **24**, 1–20.

Sugden, Roger, Robbin A. Te Velde and James R. Wilson (2002), 'Economic development and communication: problems for governance under globalisation', *L'Institute discussion paper 19*, universities of Birmingham, Ferrara and Wisconsin–Milwaukee.

Sunstein, Cass (2001). *Republic.com*, Princeton, NJ: Princeton University Press.

Te Velde, Robbin (2004), 'Schumpeter's theory of economic development revisited', in Terrence E. Brown and Jan Ulijn (eds), *Innovation, Entrepreneurship and Culture: The Interaction Between Technology, Progress and Economic Growth*, Cheltenham, UK and Northampton, MA, USA: Edward Elgar, pp. 103–123.

Te Velde, Robbin (2008), 'The dialectics of network governance: a story from the front', in Mari Jose Aranguren Querejeta, Christina Iturrioz Landart and James R. Wilson (eds), *Networks, Governance and Economic Development: Bridging Disciplinary Frontiers*, Cheltenham, UK and Northampton, MA, USA: Edward Elgar, pp. 209–25.

Walker, Thomas W. (1997), 'Introduction: historical setting and important issues', in Thomas W. Walker (ed.), *Nicaragua Without Illusions: Regime Transition and Structural Adjustment in the 1990s*, Wilmington, DE: Scholarly Resources.

Waters, Malcolm (2001), *Globalization*, 2nd edn, London: Routeledge.

Wiliamson, John (2008), 'Letter: the Spence Commission and the Washington consensus,', *Economists' Voice*, **5**(4), article 4, available at: http://www.bepress.com/ev/vol5/iss4/art4.

Wilson, James R. (2008), 'Territorial competitiveness and development policy', Orkestra working paper series in territorial competitiveness no. 2008–02, San Sebastian, Spain.

World Bank (1998), *World Development Report 1998/99*, Oxford: Oxford University Press.

Wu, Chin-tao (2002). *Privatising Culture. Corporate Art Intervention Since the 1980s*, London: Verso.

Wu, Tim (2008), *Network Neutrality FAQ*, accessed 1 September 2008 at http://timwu.org/network_neutrality.html.

Wyner, George M. and Thomas W. Malone (1996), 'Cowboys or commanders: does information technology lead to decentralization?', in J.I. DeGross, S. Jarvenpaa and S. Srinivasan (eds.) *Proceedings of the 17th International Conference on Information Systems*, Cleveland, OH: International Conference on Information Systems, pp. 63–79.

Zuurmond, Arre (1998), 'From bureaucracy to infocracy: are democratic institutions lagging behind?', in Ignace Th. M. Snellen and Wim B.H.J. van de Donk (eds), *Public Administration in an Information Age: A Handbook*, Amsterdam: IOS Press, pp. 259–72.

11. Media, governance and the public interest

J. Robert Branston and James R. Wilson

1. INTRODUCTION

When Gutenberg invented the printing press in the fifteenth century, he started a revolution that brought the communication of ideas to the masses. Since this innovation, the media industry in its various forms has progressively developed into a feature in people's lives. For example, in 2007 the British communications industry had revenues of a staggering £51.2bn (OFCOM, 2008, p. 9), equivalent to almost £850 per person in the UK.[1] Today the 'media industry' encompasses many different types of communication. Among these we can separate out as 'traditional media' the publication of newspapers/magazines and the analogue broadcast of radio and television programmes. Developed either in parallel or independently in their own right are the 'new media' of digital broadcasts and websites viewed using the Internet, themselves incorporating an increasingly sophisticated set of tools such as blogs, podcasts, e-publications and interactive features. As a society we rely on this whole range of media outlets for news and information, for political developments and discussion, for important weather or traffic updates, for warnings that may impact directly on our health and safety, and, not least, for entertainment that has the potential to bring us great welfare gains. Firms also rely on the media to inform people about their products through advertising, and it might be argued that people rely on such adverts to become informed of consumption possibilities. Taken as a whole, through all of these channels the media industry plays a key role in shaping our views and perceptions of the world.

Indeed, the influence of media on everyday life permeates widely and deeply. The media industry plays a potentially critical emancipatory role, through its direct influence on people's ability to develop knowledge and understanding around the whole range of issues impacting on socio-economic life. In this sense it impacts on both the capacity for a society to engage in critical thinking and deliberation, and the form and directions in which it does so. The media can therefore be considered a 'strategic' area of economic activity, in a similar way to the provision of electricity, water, food, transport, education or healthcare, for example. Each of these is central to modern economic life in that they fundamentally influence people's engagement in the economy more broadly.[2] Hence the organisation of activity in these industries is 'strategic' for the achievement of socio-economic welfare on the whole.

Recognition of the strategic importance of the media industry can be seen clearly in debates ongoing in the UK, for example. Here, discussions both within the media itself and among academics and policy-makers have focused on issues such as the independence of media organisations, foreign and cross-ownership, the role of the publically-funded British Broadcasting Corporation (BBC), the increasing integration of media services, and broad competition issues in the interface between media and other related markets.[3] Behind these debates there is a clear understanding that those who control the media have considerable power to influence people's views and actions (for example, voting behaviour or consumption), hence a need for great care in ensuring an appropriate structure and control of the industry.

We seek to contribute to this ongoing policy debate by analysing the media industry from a 'strategic choice' perspective (Bailey et al., 2006). In doing so, we comment on the potential for the media industry to positively or negatively impact on strategic choice processes across society. Our analysis has its origins in the strategic decision-making literature on the theory of the firm, a literature that highlights the significance of the governance of firms for achieving economic outcomes in the public interest (Branston et al., 2006a). Our argument is that in terms of achieving public interest outcomes across the economy, and more generally for the democratic and creative health of society, the governance of this strategic industry is especially critical.

The chapter is organised as follows. In section 2 we apply a strategic choice approach to analysis of the media industry, using the example of public service broadcasting to illustrate how this approach differs from a more traditional market-failure perspective. Section 3 then focuses on the extent and influence of the large media groups that are prominent in the media industry in many countries, highlighting some concerns with the governance of media activity in this context. We use the specific case of BSkyB and NewsCorp to illustrate our arguments. In section 4 we conclude with a discussion of possibilities for addressing these concerns, drawing on a number of institutional examples that represent interesting possibilities in terms of innovative forms of governance in strategic industries.

2. MEDIA AND STRATEGIC CHOICE

2.1 Contributions of the Media Industry to Strategic Choice Processes

A strategic choice perspective (Bailey et al., 2006) has been applied to analyse various specific industries and cases, including higher education (Sugden, 2004 and Wilson, in this volume), healthcare (Branston et al., 2006b) and electricity (Branston et al., 2006c). This work has its roots in Cowling and Sugden's (1987, 1994, 1998, 1999) analysis of the transnational corporation and its relationship with local economies. Cowling and Sugden built on Coase's (1937) seminal article on the theory of the firm and Zeitlin's (1974) work on corporate governance to define the transnational corporation as 'the means of co-ordinating production from one centre of strategic decision-making when this co-ordination takes a firm across national boundaries' (Cowling and Sugden, 1987, p. 12). They suggested that the concentration of strategic decision-making in the transnational corporation will result in 'strategic failure' in economies as a whole, manifested specifically in 'three sets of interrelated systematic deficiencies: transnationalism, centripetalism, and short-termism' (Cowling and Sugden, 1994, pp. 127-32). The essential argument is that 'socially efficient' outcomes are likely to be compromised in a system dominated by transnational firms. This is because the objectives of 'the few' among who key decisions are concentrated are unlikely to correspond to the objectives of 'the many' that are impacted on by firm's activities in the presence of imperfect markets.[4]

As a response to these dangers, analysis in the strategic choice tradition is centred on processes of governance; in particular, on analysis of the scope for the whole range of interests that exist within a society to feed into the strategic decisions that are made by its firms and other institutions.[5] This might be thought of more broadly in terms of the democratisation of economic processes, necessary in the context of concentrations of power within imperfect markets. The most natural departure point from Cowling and Sugden's original analysis is to analyse the governance of economic activity co-ordinated by firms. Indeed, there is a strong tradition of literature seeking to analyse the so-called 'public interest' with regards firm-based economic activity, a detailed discussion of which can be found in Branston et al. (2006a). Our focus here is on a subset of firms, those comprising the media industry. This is of particular interest because the consequences of concentration of decision-making for the public interest are potentially severe in this setting, given the influence of the media on strategic choice processes more broadly across the economy. Indeed, following Habermas (1989) the media is a critical component in cultivating the 'public sphere', that 'realm of our social life in which something approaching public opinion can be formed' (Habermas, 1964, p. 116).

The media industry is pivotal in any economy. Alongside its function in providing consumers with various forms of entertainment, the whole range of the media combine in communicating news, politics, fashions and ideas, and in contributing to knowledge generation and creativity processes. Like most economic sectors, therefore, the media has considerable potential to enhance societal welfare through what it produces. However, unlike most economic sectors it also has considerable power to shape people's thoughts, perceptions, preferences and behaviour in ways that extend far beyond the confines of the sector, and can have impacts throughout all areas of the economy.

Moreover, while the influence of the industry (or subsections of it) on people might be perceived to be harmless, or at least to be obvious and explicit to all, some of its power will be unconsidered or imperceptible to those interacting with the various media. It might be argued that the media (or subsections of it) can be an insidious influence. At worst, the media can be used as a tool to manipulate people according to the interests of the group(s) that control the industry's operations (or some other interest group with influence over those that control media output). This is seen most clearly in cases of propaganda, often in times of conflict or war, and forms the basis for Herman and Chomsky's (1988) 'propaganda model', which sets out a series of 'news filters' within privately-owned media systems that enable government and private 'elites' to control content. Such propaganda may be explicit or may take more subtle forms: Rampton and Stauber (2003), for example, present an interesting discussion of the role of 'public relations' and 'propaganda' in the US during the build-up to the currently ongoing conflict in Iraq.

Advertising is another example of the use of the media to manipulate views (Packard, 1957), with potential dangers reflected in the regulation of advertising standards and, for example, in the restriction of 'product placement' in television programmes. In particular, while it might be argued that most adults should be able to question and contrast different advertising messages, such arguments are not so clear with regards to children. Children are at early stages in forming their preferences, and this makes them particularly attractive targets for advertisers in an effort to shape preferences for life and therefore secure future consumption (see, for example, Schlosser (2002) on advertising to children in the fast food industry).

Focusing on the positive possibilities, however, it is clear that a vibrant media sector has potential to provide a powerful stimulus for open deliberation and debate, contributing to a strengthening of the basis for people's interactions with firms, government and other institutions where strategic decisions are made. From a strategic choice perspective, the development of such 'voice' as a mechanism for economic decision-making provides a fundamental balance to the often over-relied-upon mechanism of 'exit' (Hirschman, 1970). While exit as a mechanism is not inconsistent with economic democracy under certain circumstances, and is indeed an important component of choice processes in market economies, it poses significant problems if it is not balanced. First, there is not always an exit from a given circumstance; markets are characterised by imperfect competition. Second, it is a 'closed' mechanism that doesn't enable the interaction between parties that can generate creative, superior solutions.

As recognised by Hirschman (1970, p. 47), voice is not a straightforward alternative to exit: 'while exit requires nothing but a clear cut either-or decision, voice is essentially an *art* evolving in new directions'. Sugden and Wilson (2003, 2005) have argued that the ability to express voice, and to directly participate in economic decision-making, requires long-term processes of learning in a Deweyan sense: 'any social arrangement that remains vitally social, or vitally shared, is educative to those who participate in it' (Dewey, 1916, p. 7). The challenge in this sense is to create an environment in which people and their communities continually learn how to democratically express themselves, and in which the firms, governments and other organisations that make up societies learn how to develop appropriate mechanisms to facilitate and co-ordinate this input. Central to these learning processes is the ability for people to inform themselves of the relevant issues, and to interact with one-another in debating, forming and evolving their (individual and collective) views. A focus on this form of learning process is related to the analysis of creativity in economic development developed by Sacchetti and Sugden (2007, p. 18), where they 'identify the prospect of people in actual and potential publics kindling their imagination and ideas so as to shape new strategic directions in the economies in which they have an interest'. They advocate the establishment of 'public creativity forums' as *spaces* that foster free communication based on shared values. This is an idea that points also to the critical role that the media can play in facilitating virtual *spaces* for dialogue, understanding and the nurturing of creative, critical thought, as a foundation for developing the art of economic voice.

The media encompass a vital set of tools for enabling and enhancing economic participation, not only because of the nature of the sector itself but also because of the extent of public exposure to the sector. A recently published report by the British communications regulator found that the average adult in the UK spends 218 minutes per day watching television,

164 minutes listening to the radio, 24 minutes using the Internet, and 24 minutes using fixed line or mobile telephony (OFCOM, 2008, p. 28). This undoubtedly represents a considerable amount of 'social time' in people's lives. While some of this will take the form of consumption of pure entertainment, there will also be an important component of transfer of news and background knowledge, and of debate and discussion around contemporary issues.[6]

Exposure to various forms of media is thus likely to contribute significantly to processes of knowledge development among people, providing a stimulus to form and evolve views and to think and (inter-)act creatively around both individual and societal issues and problems. We suggest that a consequence of such processes is the strengthening of the capacity of people and communities to participate in the governance of different economic processes (whether this is through 'exit' or 'voice'). While the initial stimulus provided by the media forms a basis for people to interact in discussing issues that they can identify with, media tools such as blogs, online discussion forums, radio phone-ins or the 'letters to the editor' features of newspapers simultaneously facilitate such interaction. The result is a strengthening of collective consciousness, both in terms of the spectrum of different views in society and the possibilities for consensual approaches to important decisions. This, we argue, contributes to more socially efficient strategic choice processes through two channels. First, it supports the appropriate use of exit as a discrete decision-making mechanism in consumption decisions, political decisions, etc. Second, it is central to the learning processes necessary for the development of voice as a balancing mechanism to exit.

2.2 Policy and the Media Industry: Regulation and Public Service Broadcasting

The view on the significance of the media industry developed above appears broadly consistent with that taken by the British Royal Commission on the Press in 1949, as cited in McQuail (1992, p. 41):

> The democratic form of society demands of its members an active and intelligent participation in the affairs of their community, whether local or national. It assumes that they are sufficiently well-informed about the issues of the day to be able to form the broad judgments required by an election, and to maintain, between elections, the vigilance necessary in those whose governors are their servants and not their masters. . . . Democratic society, therefore, needs a clear and truthful account of events, of their background and their causes; a forum for discussion and informed criticism; and a means whereby individuals and groups can express a point of view or advocate a cause.

Once widely held, this view has been diluted or relegated behind other more dominant perspectives in the Washington Consensus inspired clamour for deregulation and free markets. In many countries the laws that once restricted the ownership and/or behaviour of the various firms that comprise the media sector are being relaxed, trusting that market forces and general competition policy will deliver outcomes that are in the public interest. For example, consider the cases of the UK and USA. In the UK, the 2003 Communications Act amongst other things relaxed many restrictions on cross-media ownership and for the first time allowed residents from outside the EEA to wholly own a British television or radio broadcaster (UK Government, 2003, sections 348 and 350). Similarly, in the USA, the Telecommunications Act of 1996 amongst other things removed limits on the number of radio stations a single entity could own nationally and significantly relaxed limits on ownership of radio stations in local markets (FCC, 2001; US Government, 1995, 1996). More recently the Federal Communications Commission (FCC), which sets media ownership rules, voted in 2007 to partially relax its ban on newspaper/broadcast cross-ownership, in effect for more than 30 years.[7]

Yet despite this trend towards deregulation there remains regulatory recognition that the media sector is special, and restrictions specific to the sector do still exist in many countries. Take, for example, the aforementioned cases of the UK and the USA. The 2003 Communications Act in the UK still mentions 'the need, in relation to every different audience in the United Kingdom or in a particular area or locality of the United Kingdom, for there to be a sufficient plurality of persons with control of the media enterprises serving that audience' (UK Government, 2003, section 375, point 2C). In the USA, FCC rules currently state that a single entity may not own a group of TV stations that reaches 40 per cent or more of all US TV households, and there are equivalent rules for the ownership concentration of local radio stations, together with limits on local radio/TV cross-ownership.[8]

Furthermore, most countries also demonstrate some appreciation of the wider importance of television and radio broadcasting by virtue of their ongoing provision for public service broadcasting, which by definition recognises that markets do not alone provide the range of outputs that society taken in its entirety might find desirable. Given this commonly held view, we now turn to consider public sector broadcasting as an exemplar of how and why the strategic failure perspective on the media diverges from traditional arguments in the economics literature, where the focus is typically on market failure. While this analysis is focused on broadcasting as a subset of the media industry, the arguments highlight issues that span the whole of the media industry.

Academic debate around public service broadcasting has been

predominantly market based and heavily influenced by Coase's analysis of the broadcasting industry in the UK and the US during the 1950s and 1960s (Coase, 1950, 1959, 1965, 1966). In particular, Coase (1966) recognised the existence of market failure in television broadcasting that was funded by advertising. The types of programmes to be developed and broadcast would be those thought to attract the size and type of audience demanded by advertisers. They would not necessarily reflect the balance of programmes that would be independently chosen by the various groups that comprise the 'public'. While this presents a rationale for public provision of broadcasting, Coase was clear that a preferable solution would be the market approach of subscription-funded television, where those wanting a particular type of programming could pay for it directly. The problem was seen as one rooted in the necessity of group provision of a service, where individual provision was technologically unviable.

This is indeed where much of the debate in recent years has focused (Cave et al., 2004; Davies et al., 1999; Armstrong, 2005; Hargreaves Heap, 2005). Technological developments in terms of satellite, cable and digital television have enabled a much wider range of channels and opened up the scope for direct provision (of specific channels or even specific programmes) to individual subscribers. Moreover, the technological landscape continues to change a pace, as faster broadband speeds in increasing numbers of homes makes Internet television increasingly popular as a convenient alternative to standard broadcasts. In the UK, for example, a significant proportion of the schedules of the BBC, ITV and Channel 4 are available to freely watch/listen online for a determined period (usually 30 days) after their initial broadcast, using Internet 'players' provided by each network.[9] The individual consumption decisions made possible by these new technologies give scope for precisely the market solution that Coase saw as preferable to public service provision in addressing the failures of advertising-funded broadcasting. These developments are thus frequently argued to negate earlier market failure arguments for public service broadcasting. Armstrong (2005, p. 283), for example, builds explicitly on Coase to argue that 'in a world where advertising funding is the sole method of commercial provision, there is a clear rationale for public policy to improve the quality and diversity of programming (as well as to diminish the nuisance of excessive advertising); but in a world where viewers can be charged directly for viewing and where spectrum is more plentiful, the benefits of substantial intervention probably outweigh its costs.' With respect to the UK case, while he concedes that there is a remaining rationale for public intervention 'linked to externality and "citizenship" concerns', he concludes that 'intervention at, or anywhere near, the current level is no longer appropriate' (2005, pp. 292–3).

The strategic failure perspective, in contrast, takes analysis beyond narrow market concerns. There are two key differences between the approaches. First, the strategic failure perspective sees each sector in relation to the development of the economy *as a whole*. As we have argued, this is particularly critical in the case of broadcasting, which has especially wide influence and can potentially enhance (or damage) processes of democratic participation in socio-economic decision-making throughout the economy. Thus failures in broadcasting have implications that extend well beyond the sector, and need to be considered and analysed as such. A market failure approach is restrictive in this sense; focusing primarily within-sector, it fails to (adequately) consider this wider context. Indeed, wider concerns are usually incorporated in a fairly ambiguous way, as an add-on along the lines of the 'externality' and 'citizenship' concerns mentioned by Armstrong (2005) for example, rather than the central subject of analysis.

The second key difference is seen in the conceptual focus of the analysis. The strategic choice framework highlights appropriate *governance of economic activity* as the means to 'efficient' development of the sector and, following our first argument, the economy as a whole. In particular, the focus is on a practical analysis of how strategic decision-making processes are actually co-ordinated among the firms and institutions that comprise the industry and the various 'publics' impacted on by eventual decisions. There is a strong emphasis on the mechanisms of voice. In contrast, market failure perspectives maintain an obsession with the mechanism of exit, leading to more distant analysis of select forces that shape the ('entry' or 'exit') decision-making environment. Thus, alongside narrow recognition of the need for government intervention in the form of public service broadcasting, they tend to focus on analysis of competition and regulation issues in improving the functioning of the market.

In the strategic choice framework competition and regulation are side issues for achieving the public interest. While it is often argued that the public interest can be achieved using appropriate forms of regulation, or if competition is sufficiently strong, these arguments are refuted in Branston et al. (2006a). They maintain that competition is rarely, if ever, an option for everyone and that regulation is insufficient because the decision-making process itself shapes people's preferences. Thus 'correct' outcomes are inherently uncertain and dynamic, while a regulatory approach typically seeks to apply fixed rules to adjust outcomes along specific lines. In the absence of democratic participation in decision-making processes we have no way of knowing what public interest outcomes would be chosen; thus distant legislation alone cannot be socially efficient. This is an argument also recognised by Coase (1966, p. 441) with specific regard to public service broadcasting: 'we cannot expect a regulatory commission to act in the public interest . . .

it must inevitably adopt certain policies and organizational forms which condition its thinking and limit the range of its policies' (p. 442).

The overall implication of applying a strategic choice framework is that without inclusive and democratic governance of decision-making within the broadcasting sector there is a risk of strategic failure. This risk is manifest both within industry (in terms of produced activities/outputs matching those that are desired), and in this case also across the economy as a whole (through the effect of broadcasting activities/outcomes on communication, information and dialogue with relevance for other areas of the economy). Of fundamental policy concern, therefore, is how decision-making within the sector is governed, regardless of whether the sector is publicly or privately owned. Thus we arrive at quite a different perspective on the rationale for public service broadcasting; broadcasting should be 'truly public' in the sense of being democratically governed by all of those with an interest. Ownership, competition and regulation issues are not the central concerns, although different settings in these regards may make it easier or more difficult to reach appropriately inclusive governance of the sector.

This illustration, applied here to the broadcasting sector, can be extended to analyse the organisation and operation of the other components of the media. Newspapers, magazines, the Internet, etc. are all strategically important for the same broad reasons as broadcasting, but their significance is frequently overlooked by conventional analysis of such sectors, whose focus is almost entirely on competition issues. The obsession with sufficient competition is no surprise in itself, given the abolition and dilution of many media-specific regulations, but such a focus is inadequate if public interest outcomes are important for the same reasons as analysed above. Indeed, it could be argued that the situation is even graver in these media sectors, because there is no equivalent to public sector broadcasting that is designed to balance the outcomes/activities determined by decision-making within powerful private firms. Having said this, it could also be argued that the low barriers to entry characteristic of new media channels (in particular the Internet) will imply over time a balancing of the power of established media groups. In the next section we turn to an analysis of the extent and influence of media groups generally, using the case of BSkyB and NewsCorp to illustrate.

3. THE EXTENT AND INFLUENCE OF MEDIA GROUPS

In many countries the media is dominated by a small number of large groups, perhaps not surprising given the aforementioned trend towards the liberalisation of media-specific legal restraints. The general consequence of these large organisations, however created, is to concentrate more and more industry decision-making power in the hands of a small number of individuals. Such concentration of power serves to exacerbate the problem of strategic failure in that much of the media is controlled by the same small group of individuals, jeopardising the plurality of views possible in a more diversely owned media sector. Indeed, companies that are part of the same group habitually provide content for one another, thereby reducing diversity and exposure to different issues and perspectives. While a perception of choice is maintained, the viewer/listener/reader is in fact getting a variation on the same set of messages from multiple media avenues. It is also likely that the consumers are unaware of the common control of the different parts of these media groups, so they can't be alert to the possible dangers this implies. In short, the concentration of production among a few large media groups gives the elite group of people that control these corporations inordinate influence over the communities in the societies where they operate. Taking an illustration from the UK, we now consider the case of the broadcasting group BSkyB.

BSkyB is a FTSE 100 publicly traded company, and is the largest commercial broadcaster in the UK and Ireland, dominating the market for subscription funded television through its own satellite broadcast system, and by making its channels available through other platforms such as cable or the terrestrial digital service 'Freeview'. In the year ended 30 June 2007, BSkyB had 9.8 million homes (40 per cent of television homes) subscribing to its television services, which helped generate revenues of £4551 million and operating profit of £815 million (BSkyB, 2007, p. 52). BSkyB is the dominant commercial TV broadcaster in the UK and Ireland, offering subscription television, internet, telephony and betting services, and also owns subsidiary companies that print magazines and make satellite receiving equipment (BSkyB, 2007, p. 81). Moreover, in November 2006 BSkyB purchased a 17.9 per cent share in ITV, the major non-subscription commercial TV company in the UK (BSkyB, 2007, p. 67), although this was mandated to be reduced to below 7.5 per cent following a Competition Commission investigation, a decision that is currently being appealed (BERR, 2008).[10]

The largest shareholder of BSkyB is the US-based News Corporation (Newscorp), which owned more than 39 per cent of the shares as of July 2007, giving it effective control of the organisation since the next largest shareholder had less than 6 per cent (BSkyB, 2007, p. 103). This large transnational corporation has numerous significant interests in the global media industry, its influence spanning every continent. Firms controlled by Newscorp include Fox Entertainment Group, Twentieth Century

Fox Film and the *New York Post* in North America, Sky Latin America, Star TV in Asia and the Middle East, and a large array of newspaper titles in Australia. Within the UK, alongside BSkyB Newscorp controls several of the most important national newspaper titles (*The Times*, *The Sunday Times*, *The Sun*, *The News of the World*), a significant book publisher (HarperCollins), and has a holding in The Wireless Group radio network.[11] Furthermore, its influence is also felt significantly through its American interests in film and television which sell large amounts of content into the UK market. In terms of presence in new media, Newscorp also owns the popular MySpace website.[12] Indeed, Newscorp is a company that wields enormous global power; both economically (in the year to 30 June 2008 it had revenues of US$32.9bn and profits of nearly US$5.4bn) and in terms of influence.

Given the scope of operations of BSkyB and Newscorp, there are severe dangers in terms of strategic failure. Imagine an individual who receives their TV, Internet and telephone services from BSkyB, purchases books and newspapers from Newscorp subsidiaries, and also uses the MySpace website for social networking and channelling online content. Furthermore, perhaps a significant part of the media content provided by BSkyB and other Newscorp subsidiaries to this person originates from other Newscorp controlled companies, such as Star TV, Fox Entertainment or Twentieth Century Fox Film. It is not difficult to envisage a scenario in which almost the entirety of the news, information, debate and entertainment to which that person is exposed is essentially controlled by Rupert Murdoch (Chairman and Chief Executive Officer) and the other narrow groups in position to exert a controlling influence over BSkyB and Newscorp.

Given the extent of exposure to these services, this person cannot help but be influenced by the agendas of those who control the content of these media outlets. Influence may be as subtle as reporting of particular news stories in a certain tone, under- or over-featuring of a specific line of news, or systematic selection of commentators with certain views to provide opinion and discussion features. Moreover, given the plurality of titles and brands it is likely that this person is unaware of the extent to which they are reliant upon one media empire, their knowledge and views shaped potentially by the perspective of just one person. Indeed, an example which serves to illustrate the (hidden) power of such media groups can be found in Rupert Murdoch's admission to a UK Parliamentary enquiry committee that he personally decides the political line of the *The Sun* and *The News of the World*, two of the UK's biggest daily and Sunday newspapers respectively (each with a 58 per cent market share according to Newscorp (2008, p. 20)). He also noted that he chose himself which political parties they would back during election time.[13] Thus one man's opinions and views, unaccountable to anyone (since he controls the Newscorp organisation), exercises considerable influence over the public perception of politics in the press and – given the extent of the Newscorp's and BSkyB's interests – probably also through other media channels.

We have used perhaps the most obvious example to illustrate the extent and influence of large media groups. However Newscorp is by no means the only large entity in what is a highly concentrated and interdependent set of activities, certainly at national and increasingly at global level (as the Newscorp example illustrates). This reflects, for example, the economies of scale in comprehensive news collection and in the marketing and diffusion of media content. Moreover, while it might be argued that the new media have lowered barriers to entry, concentration remains the norm here too. This is evidenced by the dominance of the likes of MySpace and Facebook (social networking and viral communication of news/information/entertainment), Yahoo!, MSN and Google (email, search and multiple news/information/entertainment-related applications), and the BBC (news/information/entertainment and comment/analysis), whose website is consistently in the top ten most visited in the UK and top 40 most visited English language sites globally (see www.alexa.com).

Moreover, in many national settings, private-sector concentration in the media is complemented with an additional significant 'public-service' media presence (in particular in television and radio, but also in other media spheres as the BBC case illustrates). Theoretically, concentration of activities in a public setting raises exactly the same issues from a strategic failure perspective; the only difference is that a distinct set of interests is able to express their influence. In either case, the absence of appropriately balanced voice and exit mechanisms of governance is likely to imply strategic failure, both within the industry and impacting throughout the economy. In this case the broader impact is seen in a failure to expose people to a sufficiently broad set of issues, presentations and perspectives; those that dominate among a concentrated media are determined by a narrow group of individuals. This failure fundamentally undermines societal capacity for critical thought and engagement in processes of deliberation, which we have suggested are central to efficient processes of socio-economic development.

Moreover, not everyone in a particular society will be affected uniformly by this failure. Certain groups, for example pensioners, might be uncommonly reliant upon a small number of media outlets, which would make them especially vulnerable. For example, Ofcom reports that 'television is the main source for UK and world news and entertainment for nearly 70 per cent of consumers' in the UK (Ofcom, 2008, p. 15), with the over-65s spending 83 more minutes a day watching television than the UK average (Ofcom, 2008, p. 11). Furthermore, 69 per cent 'of those aged 65-74 identify television as the media activity they would miss most . . . rising to 77 per cent among the over 75s' (Ofcom, 2008, p. 11), all suggesting that this group is uncommonly reliant upon television compared with other groups within society. If the television channels that they receive are themselves particularly subject to narrow, unbalanced governance, then the strategic failure impact is magnified among this uncommonly exposed subset of society. Similarly, consumers who buy bundled services (television, telephone, mobile phone, Internet) from one company, which then becomes a dominant or even sole channeller of media outputs into their household, may be particularly vulnerable. This is clearly a concern in the UK, for example, where 40 per cent of households now buy communications in bundled packages (Ofcom, 2008, p. 1).

4. CONCLUSIONS: WHAT TO DO

Application of the strategic failure perspective gives us unique insight into the significance of the media sector for the overall development of the economy. We have argued that this is an especially important industry for a democratic society that is seeking to develop its economy according to the aims of its people. The influence that the media potentially has over people's ideas, perceptions and actions make the sector particularly susceptible to strategic failure: a situation where development of the economy proceeds in line with the exclusive interests of a (number of) small, powerful group(s). On the other hand, the media can potentially provide a crucial stimulus for the interaction, discussion and debate among people that underpins the economic democracy required to overcome strategic failure. We suggest that the critical factor distinguishing these two scenarios is the governance of the sector itself, and in particular the determination of strategic decisions. This differs markedly from most existing economic debate around the media as an industry, where the focus tends to be on issues of market failure and competition policy. We suggest here that regulation and competition issues should not be the guiding concerns. While they can play an important role in enhancing governance through impacting on the mechanism of 'exit', they frequently neglect the more difficult issue of developing the mechanism of 'voice'.

We have also pointed here to a situation in which there has been in many places a withdrawal of specific regulations designed in recognition of the media's strategic role in exposing society to news, information and debate that condition, among other things, political democracy. A commonly held perception has been that the media industry should operate in the context of normal market forces, a view often argued with reference to new technologies reducing previously acknowledged market failures. Parallels could be drawn in this sense with the financial sector, whose current crisis should be put in the context of a sustained period in which arguments were made for progressive reductions of specific regulations without a rigorous analysis of whether or not the rationale for checks and balances still existed in a changing technological and market environment. However our central argument, in both cases, would suggest that focusing on regulatory solutions alone misses the point: the appropriate development of voice is paramount if the governance of such strategic sectors is to accurately reflect the public interest.

Essentially this analysis calls for a democratisation of the decision-making processes of the various institutions (public or private) that constitute the media industry, so as to ensure the articulation of voice. While this is undoubtedly a long-term process that itself requires significant learning, there are existing institutional examples on which we might draw. Within the media sector itself, for example, we might look to the BBC. This is a large and successful media organisation with an interesting and arguably unique governance structure, a detailed analysis of which can be found in Branston and Wilson (2008). Neither privately nor publically controlled in the conventional (government) sense, the BBC has no shareholders and is explicitly mandated by Royal Charter to operate in the interests of the 'public' of licence-fee payers that it is tasked to serve. To carry out this task the Charter establishes a BBC Trust, which is required 'actively to seek the views of, and engage with, licence fee payers' (DCMS, 2006: Section 26). This operates alongside a series of mechanisms through which public 'voice' can be developed, including 'National Broadcasting Councils' for different parts of the UK, the geographical dispersion of production and 'drop-in' facilities, and considerable website space dedicated to informing on its activities and enabling comment and feedback (Branston and Wilson, 2008).

Similarly, Network Rail, the British railway infrastructure company, has no shareholders and a mission 'to deliver a safe, efficient, sustainable and reliable rail service' (Network Rail, 2007/08, p. 1), rather than profits. Network Rail owns and operates the railway infrastructure in Great Britain, providing the facilities (such as track and stations) for the train operating companies to use when providing services to the public or to industry.[14] Legally it is a company limited by guarantee, which means that it is a private limited company. However rather than shareholders, it has 108 members (Network Rail, 2008, p. 20) who provide the corporate governance and hold the management team to account. The company is professionally run by a board of directors and senior management team with the aim of making profits, as with a traditional shareholder owned firm. However, instead of paying dividends to shareholders, profits are re-invested in the industry.

Such a model is potentially more likely to avoid problems of strategic failure because the organisation is explicitly designed to make decisions informed by the 'many' with an interest in the railways, rather than the 'few' with an interest in their private return. Our arguments would suggest nevertheless that its success in doing so will depend on the governance mechanisms in place to articulate these interests. In the case of Network Rail mechanisms are focused on the role of its 'members', who include representatives from all stakeholders in the industry. Members are appointed following an open application process and appointment panel, and there are two main categories: Industry Members and Public Members. Industry Members are drawn from all of those firms which operate within the railway industry; from firms that operate trains, to companies that provide engineering services to Network Rail. Public Members are drawn from all other individuals or organisations which have an interest in how the railway functions; from individual members of the public with an uncommon interest in the sector, to worker representatives, to regional transport organisations. Thus there is an open opportunity for everyone with an interest in the company to engage with its decision-making process, either directly through becoming a member and holding management to account, or indirectly through representations to those selected to be members.[15]

Whilst different in their nature, both of these examples illustrate institutional possibilities that offer significant potential for the development of governance solutions that can reduce the dangers of strategic failure in key economic sectors. We suggest that exploring, testing and advancing these possibilities in the context of the media industry is an important challenge for public policy. Achieving public interest outcomes across the economy relies on the development of capacity for critical thought, debate and interaction among the people and institutions that comprise a society. The media has a critical role to play in facilitating this: in terms of accurate, factual reporting of news and other information; in terms of exposure to a broad diversity of different views and perspectives around topical issues; and in terms of facilitating forums that serve as a stimulus to debate and interaction. Ensuring that the media is able to impact positively on enhancing governance processes throughout the economy requires that policy-makers pay specific attention to developing the evolution of appropriate governance mechanisms within the media industry itself.

NOTES

1. This refers to the communications industry as based on the elements monitored by OFCOM, which include TV, radio, telephony, and the Internet. Including other sectors such as newspapers and magazines would generate a significantly higher figure. For example, as of August 2008 annual consumer expenditure on magazines in the UK was £1.64 billion (see 'Country lifestyle magazines thrive as credit crunch hits the lads' mags', *The Times*, 15 August 2008, p. 51).
2. To take one example, without appropriate access to electricity it is difficult to gain access to a whole range of welfare enhancing products (refrigerators, washing machines, televisions, etc.) and so developments in this industry might be seen as disproportionately important to the economy.
3. See, for example: DCMS (2005, 2006); Armstrong (2005); Hargreaves Heap (2005); Cave et al. (2004); UK Government (2003); and Vickers (2002).
4. Their analysis is strongly related to Hymer's (1972) theory of uneven development, by which he identified a '*correspondence principle* relating centralization of control within the corporation to centralization of control within the international economy' (p. 123). More recently such a perspective has been applied by Sugden and Wilson to a critique of mainstream development policy and practice under the so-called Washington consensus (2002), and to an analysis of current globalisation (2005).
5. 'Strategic decisions' refer specifically to those that are at the top of the decision-making hierarchy. See Pitelis and Sugden (1986) for a discussion of different types of decision. See also Bailey et al. (2006) for a discussion of how the concept of 'governance' relates to the strategic choice perspective.
6. Indeed, the boundaries between 'pure entertainment' and 'news, background knowledge and topical discussion' are not so clear. Many programmes are likely to bridge these boundaries, with issues featured in dramas or soap operas, for example, corresponding to key contemporary issues, and thus providing stimulus for thought, discussion and formation of views.
7. The vote to relax the ban on cross-ownership is being challenged in court. See http://www.fcc.gov/cgb/consumerfacts/reviewrules.html for more detail on the FCC and its ownership rules, accessed on 9 September 2008.
8. See http://www.fcc.gov/cgb/consumerfacts/reviewrules.html, accessed on 9 September 2008.
9. A million users reportedly downloaded 3.5 million TV programmes in the two weeks following the launch of the BBC's iPlayer, (see 'What is the true impact of the BBC iPlayer?', *The Guardian*, 11 February 2008).
10. It is reported that Italian media group, Mediaset, is considering making a bid for the BSkyB stake in ITV and possibly even a full take-over bid for the company. Should this take place it would raise a different set of strategic failure concerns given that Mediaset is the dominant commercial broadcaster in Italy and has a reputation of providing core programming consisting of 'soap operas, game shows, reality shows and near-naked dancing girls' (see 'ITV in the sights as Berlusconi's empire eyes fresh conquests', *The Times*, 12 September 2008).
11. See http://www.newscorp.com/ and Newscorp (2008) for a full listing of all of the worldwide holdings of Newscorp.
12. MySpace, one of the fastest growing websites in history, is a path-breaking social networking site thought to have more than 110 million monthly active users around the world as of January 2008. See http://www.web-strategist.com/blog/2008/01/09/social-network-stats-facebook-myspace-reunion-jan-2008/.
13. See http://news.bbc.co.uk/1/hi/uk/7110532.stm, accessed on 11 September 2008.
14. Network rail doesn't operate passenger or freight trains, nor own the rolling stock used to provide these services. These are provided by train operating companies, freight operating companies and rolling stock companies respectively.

15. J. Robert Branston, one of the authors of this chapter, was a Public Member of Network Rail for five years up to November 2007.

REFERENCES

Armstrong, M. (2005), 'Public service broadcasting', *Fiscal Studies*, **26** (3), 281–99.

Bailey, D., L. De Propris, R. Sugden and J.R. Wilson (forthcoming), 'Public policy for European economic competitiveness: an analytical framework and a research agenda', *International Review of Applied Economics*.

Branston, J.R. and J.R. Wilson (2008), 'Transmitting democracy: broadcasting policy and the BBC', mimeo, University of Bath.

Branston, J.R., K. Cowling and R. Sugden (2006a), 'Corporate governance and the public interest', *International Review of Applied Economics*, **20** (2), 189–212.

Branston, J.R., L. Rubini, R. Sugden and J.R. Wilson (2006b), 'The healthy development of economies: a strategic framework for competitiveness in the health industry', *Review of Social Economy*, **LXIV** (3), 301–29.

Branston, J.R., R. Sugden, P. Valdez and J.R. Wilson (2006c), 'Creating participation and democracy: electricity reform in Mexico', *International Review of Applied Economics*, **20** (1), 47–68.

British Sky Broadcasting (BSkyB) (2007), *Annual Report*, accessed 18 March 2008 at http://phx.corporate-ir.net/phoenix.zhtml?c=104016&p=irol-reportsannual.

Cave, M., R. Collins and P. Crowther (2004), 'Regulating the BBC', *Telecommunications Policy*, **28**, 249–72.

Coase R.H. (1937), 'The nature of the firm', *Economica*, **4** (16), 386–405.

Coase, R.H. (1950), *British Broadcasting: A Study of Monopoly*, London: Longman.

Coase, R.H. (1959), 'The Federal Communications Commission', *Journal of Law and Economics*, **2**, 1–40.

Coase, R.H. (1965), 'Evaluation of public policy relating to radio and television broadcasting: social and economic issues', *Land Economics*, **XLI** (2), 161–7.

Coase, R.H. (1966), 'The economics of broadcasting and public policy', *American Economic Review*, **56** (1/2), 440–7.

Cowling, K. and R. Sugden (1987), *Transnational Monopoly Capitalism*, Brighton: Wheatsheaf Books.

Cowling, K. and R. Sugden (1994), *Beyond Capitalism: Towards a New World Economic Order*, London: Pinter Publishers.

Cowling, K. and R. Sugden (1998), 'The essence of the modern corporation: markets, strategic decision-making and the theory of the firm', *The Manchester School*, **66** (1), 59–86.

Cowling, K. and R. Sugden (1999), 'The wealth of localities, regions and nations: developing multinational economies', *New Political Economy*, **4** (3), 361–78.

Department for Business Enterprise and Regulatory Reform (BERR) (2008), 'Final decisions by the Secretary of State for Business, Enterprise and Regulatory Reform on British Sky Broadcasting Group's acquisition of a 17.9 per cent shareholding in ITV plc dated 29 January 2008', accessed 20 June at www.competition-commission.org.uk/inquiries/ref2007/itv/pdf/sky_berr_decision.pdf.

Department for Culture, Media and Sport (DCMS) (2005), *Review of the BBC's*

Royal Charter: A Strong BBC, Independent of Government, Green Paper, March 2005, London: Stationery Office.
DCMS (2006), *A Public Service for All: The BBC in the Digital Age*, White Paper, March 2006, London: Stationery Office.
Dewey, John (1916), *Democracy and Education*, New York: Macmillan.
Federal Communications Commission (FCC) (2001) *Review of the Radio Industry*, accessed 9 September 2008 at www.fcc.gov/mb/policy/radio.html.
Habermas, J. ([1964] 1997), 'The public sphere', reprinted in Peter Golding and Graham Murdock (eds), *The Political Economy of the Media, Volume II*, Cheltenham, UK and Lyme, USA: Edward Elgar.
Habermas, J. (1989), *The Structural Transformation of the Public Sphere*, Cambridge, MA: MIT Press.
Hargreaves Heap, S.P. (2005), 'Public service broadcasting', *Economic Policy*, January, 111–57.
Herman, E.S. and N. Chomsky (1988), *Manufacturing Consent: The Political Economy of the Mass Media*, reprinted 2002, New York: Pantheon Books.
Hirschman, A.O. (1970), *Exit, Voice and Loyalty: Responses to Decline in Firms, Organizations and States*, Cambridge, MA: Harvard University Press.
Hymer, S. (1972), 'The multinational corporation and the law of uneven development', in: Jagdish N. Bhagwati (ed.), *Economics and World Order: From the 1970s to the 1990s*, London: Macmillan.
Kaufman, G. et al. (1999), *Third Report: The Funding of the BBC*, report of the Culture, Media and Sport Committee, December 1999, London, accessed at www.publications.parliament.uk/pa/cm199900/cmselect/cmcumeds/cmcumeds.htm.
McQuail, D. (1992), *Media Performance: Mass Consumption and the Public Interest*, London: Sage.
Network Rail (2007/2008), *Corporate Responsibility Report 2007/08*, accessed 23 September 2008 at www.networkrail.co.uk/.
Network Rail (2008), *Network Rail Ltd Annual Report and Accounts 2008*, accessed 23 September at www.networkrail.co.uk/.
Newscorp (News Corporation) (2008), *Newscorp Annual Report 2008*, accessed 19 September at www.newscorp.com/investor/annual_reports.html.
Office of Communications (Ofcom) (2008), *Communications Market Report 2008*, London, accessed 15 August at www.ofcom.org.uk/research/cm/cmr08/.
Packard, V. (1957), *The Hidden Persuaders*, reprinted 1991, London: Penguin.
Pitelis, C.N. and R. Sugden (1986), 'The separation of ownership and control in the theory of the firm', *International Journal of Industrial Organization*, **4** (1), 69–86.
Rampton, S. and J. Stauber (2003), *Weapons of Mass Deception: The Uses of Propaganda in Bush's War on Iraq*, London: Robinson.
Sacchetti, S. and R. Sugden (2007), 'Creativity in economic development: space in an inferno', Institute for Economic Development Policy discussion paper 2007-02, University of Birmingham.
Schlosser, E. (2002), *Fast Food Nation: What the All-American Meal is Doing to the World*, London: Penguin Books.
Sugden, R. (2004), 'A small firm approach to the internationalisation of universities: a multinational perspective', *Higher Education Quarterly*, **58** (2/3), 114–35.
Sugden, R. and J.R. Wilson (2002), 'Development in the shadow of the consensus: a strategic decision-making approach', *Contributions to Political Economy*, **21**, 111–34.

Sugden, Roger and James R. Wilson (2003), 'Urban and regional prosperity in a "globalised", "new" economy: an agenda and perspective', in Roger Sugden, Rita Hartung Cheng and Richard Meadows (eds), *Urban and Regional Prosperity in a Globalised, New Economy*, Cheltenham, UK and Northampton, MA, USA: Edward Elgar.

Sugden, R. and J.R. Wilson (2005), 'Economic globalisation: dialectics, conceptualisation and choice', *Contributions to Political Economy*, **24**, 1–20.

UK Government (2003), *The Communications Act 2003 (c.21)*, London, accessed at www.opsi.gov.uk/acts/acts2003/20030021.htm.

US Government (1995), *Senate Report 104-023 – Telecommunications competition*, accessed 9 September 2008 at http://thomas.loc.gov/cgi-bin/cpquery/R?cp104:FLD010:@1(sr023).

US Government (1996), *Telecommunications Act of 1996*, accessed 9 September 2008 at http://frwebgate.access.gpo.gov/cgi-bin/getdoc.cgi?dbname=104_cong_public_laws&docid=f:publ104.104.

Vickers, J. (2002), 'Competition policy and broadcasting', speech at the IEA Conference on The Future of Broadcasting, 24 June, accessed at www.oft.gov.uk/News/Speeches+and+articles/2002/index.htm.

Zeitlin, M. (1974), 'Corporate ownership and control: the large corporation and the capitalist class', *American Journal of Sociology*, **79** (5), 1073–119.

12. Quantity, quality and creativity

Francesco Sacchetti

In social sciences it is possible to distinguish two research approaches that are widely debated. The first one conceives the world along a number of variables and aims at analysing the association between them; the second one is devoted to the holistic comprehension of specific social realities. This obviously lays the foundations for debate between quantitative and qualitative approach.[1]

In the brief space of this chapter I will only touch on some of the concepts related to the above-mentioned debate. The specific purpose here is to link the research process in the social sciences with the dimension of creativity. Rather than addressing creativity as an ordering property of chaos – as it happens in a number or contributions on cosmogony (Arlow, 1982; Bonnefoy, 1992) – or as an exclusive property of genius often associated to dissolute living and eccentricity, I look at creativity as a property belonging to every single human being (Chomsky, 1964; Amabile, 2001).

The notion of creativity is linked to a human being's property that is often not well defined. In this chapter I will use two meanings of creativity: one comes from the Latin word *creàre*, which subtends to the idea of creation from nothing, as suggested by the Jewish–Christian tradition, where the divine accomplishes creation in a linear way (it had a beginning and an end). The other one comes from the Greek word *kraínó*, meaning to produce or to complete, from which derives another word, *krantòr*, which means dominator, and *kreiòn*, the one who makes, the one who creates. In the Greek etymology the action of creation is strictly connected to the domination of something that allows creation; nevertheless Greeks had a circular conception of motion structures and time (Gobo and Tota, 1994), that suggests the idea of transformation instead of a creation from nothing.

The etymologies of the word help us to understand why Chomsky (1964) speaks about creativity instead of productivity when referring to the human being's ability to formulate and to understand new sentences in accord with a grammar. He distinguishes between *rule-changing creativity* and *rule-governed creativity* (ibid., p. 22). The distinction pertains to linguistic production regarding grammar rules: an individual could be creative by means of changing the language in some ways (first type), or because although he[2] does not modify the language, he rather understands it and produces new sentences (second type). In the first case the property concerns a creation that breaks with what has been up to that particular moment. It is an act that bypasses established rules and constitutes something original, which did not exist before in that ambit (which reminds us of the Latin meaning). I will link this form of creativity to the qualitative approach. In contrast, I will refer to *rule-governed creativity* (linking it back to the Greek meaning) in relation with the quantitative approach; it is an application of rules that come from both substantive and procedural knowledge.

In the next sections I suggest in what respect the Chomskian dyad can help to identify the creative aspect of research as a constituent and normal part of the knowledge production process. Contextually, I will propose creativity as being a property of both quantitative and qualitative research approaches: the substantial differences, I argue, regard when, in what phase of knowledge production creativity is particularly necessary, and how it is expressed. I then apply my considerations to the different treatments of creativity in research, looking at the plurality of structures that research design may assume, reflecting the researcher's decision to adopt a quantitative or qualitative approach.

THE QUANTITATIVE APPROACH

How does the researcher observe reality? Should he uses preconstituted tools, built up on his previous knowledge (*etic* perspective), or rather leave his attention free to move on to what people feel relevant for their own life (*emic* perspective).[3] To understand how this becomes a choice that influences the entire set up of the social scientist's work, I briefly introduce the concept of 'data collection scheme'. The latter has been defined as 'the whole of linguistic categories by which we examine the world' (Bruschi, 1999, p. 327, author's translation). To the data collection scheme (interview outline or observational protocol) are referred several structuring levels, according to the researcher's freedom to address his attention (while he is observing) to elements that he considers interesting and relevant for realising his research. The scheme could be (1) completely structured, (2) partially structured, (3) not structured.

Bruschi (1999, p. 327) defines structuring as a continuum along which it is possible to set up the observational tools:

On the left-hand extreme of the continuum there are those tools[4] by

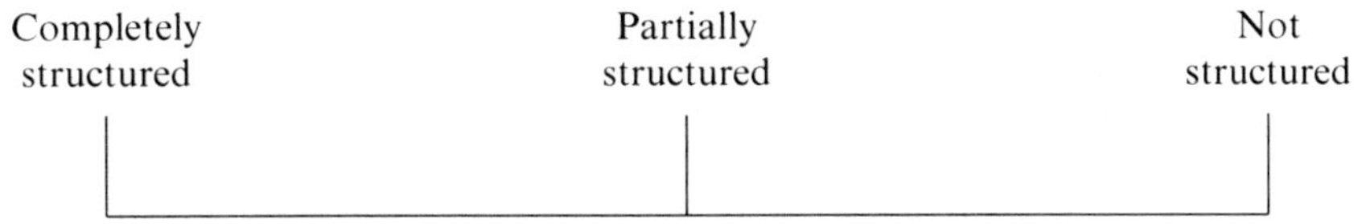

Figure 12.1 Structuring as a continuum

which, in the design phase, one defines clearly and unambiguously, what to observe. This allows a comfortable comparison among different observations. Moving to the right-hand side of the continuum, the degree of freedom with which the researcher observes, increases. So the researcher will be free to collect every event he considers interesting in order to gain information.

I would also underline that to define a data collection scheme with different degrees of rigidity depends on the establishment of more or less constraining rules that the researcher should follow. For example, in a completely structured data collection scheme the researcher would indicate what to observe, how to observe it and how to record it. This is generally the case in which the researcher adopts a quantitative approach, which would be effective once one disposes of a satisfying knowledge of the research object. In this case, from a research-design point of view, when the object studied and the theory are defined, the researcher would formulate the hypothesis (that will be checked during data analysis)[5] as well as the model specification. He proceeds then to set up a sampling plan – which ideally should be random in order to make inferential procedures on research results possible – on the basis of those proprieties which have to be representative. After these steps, he operationally defines all the properties of the object studied. This consists in giving a set of rules that allow him to transform the attributes of an object into the modalities of a variable. Data collection and matrix data entering would be followed by data analysis, checking of the hypothesis and communication of the results.

According to the quantitative approach, the researcher can operationally define properties in order to build up a set of variables and, subsequently, organize data in a matrix. This process is consistent with the specification of the model, which includes both its properties, as well as the hypothesized relationships between them.

This can be identified as the first creative phase of the quantitative approach: relying on the strength of the knowledge acquired about the problem, the researcher defines which properties are relevant for his aims. He then creates a model which takes into account both a vision of reality and formal aspects of definition, depending on the purpose of his research.

This ability to move deftly between reality and the necessary reduction of complexity inside a model needs particular cognitive effort: it requires the application of theoretical knowledge about relationships between properties in a model that presupposes a cognition, or at least a hypothesis, about the direction of relationships.

Once the model has been specified, the researcher enters in the second creative phase; in this case properties are defined in a more operative way. According to Marradi (2007, p. 107) an operational definition is 'the set of rules and conventions that allow to transform a property into a variable inside a matrix' (author's translation). He also suggests that those conventions regard not only every single property, but also every new research. Especially in social sciences, where shared and inter-subjective definitions are not so common, the researcher has a wide margin to express his creativity despite the fact of working with an approach (the quantitative) that is historically considered objective and characterized by a low level of influence of the knowing subject.

For example, the use of different types of scale (for example a Likert scale rather than a feeling thermometer) to collect states on a property, rather than a more or less sensitive plan of codification,[6] are all due to the choice mediated by the researcher's creativity, when he applies his competences to a new object of enquiry. This vision of the process does not break the orderliness claim of the quantitative approach, but nor does it underestimate the creative role of the researcher during the process of model specification and operational definition. The latter impacts on the whole research and reflects the researcher's creative effort as an expression of what I would name, following Chomsky (1964), *rule-governed creativity*.

Building on Chomsky (ibid.), I propose an interpretation of research according to which the creative act supposes both substantial and procedural competences, which allows us to operatively define properties and to transform them into variables. This type of creativity is strongly related to the common practice of problem solving based on the logic of learned rules. Here the warning of *kramtòr* (the dominator) echoes strongly: in this case the dominion concerns what is known, that means the capacity to describe and question it in an appropriate way depending on previous knowledge (*etic* perspective). The creative phase concretely leads, on the one hand, to the construction of the collection tool, which takes place before the gathering of information. On the other hand, it offers the advantage of proposing a testable hypothesis, as previously defined variables are analysed. The representation of reality – and its correspondence to the collected information – is the major issue here, because the creative effort must work within the established rules. Conformity to existing rules allows us to deconstruct results and build upon them once the original research tool has run out of its collection function.

Non-Structured Tools

It is not always possible to think of a particular phenomenon and objects a priori: for example, the relevant 'codified knowledge' (Polanyi, 1958) could be unavailable, available only partially, or obsolete (Cowan et al., 2000). In these cases, model specification is impossible or inoperative, following of the quantitative approach criteria. According to Nigris (2003) such an outcome could even be the conscious choice of not structuring when one recognises:

> the excess of what is knowable over what is known, mainly by avoiding the supposition of holding right at the beginning the whole set of concepts concerning the description of the observed world. (2003, p. 42, author's translation)

Starting from these assumptions, the use of a highly structured tool is not always a viable research strategy. By eventually overestimating the available knowledge, the researcher runs the risk of behaving – to recall an old Jewish saying – 'like the nocturnal drunkard who searches for his lost keys under a street light because that is where they are easier to see' (Piore and Sabel, 1984, p. 222). The light of the 'quantitative streetlamp' is bright but also strictly focused on that specific part of reality on which the researcher directed it. All the things that fall outside the light run the risk of being ignored, making the research useless.

Differently, in an unstructured data collection scheme the researcher tries to be led by people's narration, and therefore leaves them free to picture the world at their level of comprehension. In this context the researcher has to find the way that he considers the most appropriate. Because of the complexity of reality and the narrowness of individuals' knowledge (Hayek, 1945), in some cases it is not possible to acquire previous knowledge about the object of research. This is, for instance, the case of local phenomena, whose knowledge has never been codified in any form. Nevertheless a low structuring tool is advisable when the object reflects issues that are difficult to observe, such as personal relationships or, more critically, people's point of view (*emic* perspective). The whole strand of qualitative studies comes from considerations of that kind and finds its application building on the assumption that an individual is able to argue about his everyday life. The researcher's task is then to arrange a low structured research tool, one that would leave to the speaker the possibility to express himself with his own words and at his level of comprehension (Montesperelli, 1998, p. 66).

The aim is to recognize in the subject a better awareness of his own environment, as compared to the researcher's. For example, a community that decides to live in a radically different way from the rest of society is hard to investigate a priori, not least because of the poor contextual knowledge that the researcher has of that world. For these reasons considering, for example, the technique of in-depth interview, it will be desirable that the subjects lead the trajectory and the unfolding of interviews. The researcher steps in only when the interview is protracted in some uninteresting ways that are recognized as being useless for the research aims. This allows, especially in preliminary steps, to use, or to fix, 'sensitizing concepts' around which to focus the attention during later interviews, other than introducing the researcher to a new way of thinking, which is different from his own and from his mental categories. The expression 'sensitizing concepts' was coined by Blumer (1969), by which he means those concepts that prepare the researcher to perception, to follow some ways instead of some others, that focus his attention on specific issues or situations during the development of the research:

> I think that thoughtful study shows conclusively that the concepts of our discipline are fundamentally sensitizing instruments. Hence, I call them 'sensitizing concepts' and put them in contrast with definitive concepts such as I have been referring to in the foregoing discussion. A definitive concept refers precisely to what is common to a class of objects, by the aid of a clear definition in terms of attributes or fixed bench marks. This definition, or the bench marks, serves as a means of clearly identifying the individual instance of the class and the make-up of that instance that is covered by the concept. A sensitizing concept lacks such specification of attributes or bench marks and consequently it does not enable the user to move directly to the instance and its relevant content. Instead, it gives the user a general sense of reference and guidance in approaching empirical instances. Whereas definitive concepts provide prescriptions of what to see, sensitizing concepts merely suggest directions along which to look. . . . They lack precise reference and have no bench marks which allow a clean-cut identification of a specific instance and of its content. Instead, they rest on a general sense of what is relevant. There can scarcely be any dispute over this characterization. (ibid., pp. 149–50)

Starting from sensitizing instruments, the level of structuring of the scheme can change during the research development but, as Cozzi and Nigris (2003, p. 265) notice, it can only get higher. At the beginning of the research, if the study object is not well known, the observer lets events lead his involvement. Differently, at the end of the explorative work when he knows more about the object of his study, if necessary, he might decide to adopt a more structured data collection scheme. Having gone through these stages, *the researcher can examine the world through learned categories, or can build a model based on learned knowledge*, therefore being in a position to undertake simulations.[7]

THE QUALITATIVE APPROACH

When the researcher adopts a quantitative approach, he moves inside a complex set of rules with which he is confronted when using his creativity. Differently, the qualitative approach is not based on a set of procedures or necessary and sequential steps that the researcher has to follow to set up and use his own gathering tools. Moving from the assumption that knowledge is not adequate to describe a certain portion of reality, what qualitative research really tackles is the cognitive effort to understand mental categories of other subjects and how they see their world.

The key driver of qualitative approaches remains on an implicit level: whatever the reason for its not being possible to have sufficient knowledge, the researcher has to tackle the issues of inaccessibility of information and of lack of comprehension when designing research. The qualitative approach allows us to break those impasses. By entering into the field, the researcher interacts, experiences directly how people live, what their practices and habits of thought are, as well as what mental categories they use to give sense to reality. This allows him to define problems on the strength of a new and pertinent knowledge that does not belong to people who live in an external context.

Differently, whilst a quantitative approach requires using procedural and substantial competences, I suggest that in a qualitative enquiry the researcher's fieldwork is considered as a 'performance' because of its adaptive character. The researcher is constantly confronted with unforeseen situations, surrounded by an unknown environment. Also, this use of the notion of 'performance' comprehends both elements of the process as well as of the outcome of fieldwork, as it recalls peculiar characteristics of qualitative work: action and interaction, personal involvement and, above all, orientation to a purpose (in this context, the teleological purpose of knowledge production). According to Victor Turner (1982, p. 91):

> *Performance* . . . is derived from the Middle English *parfournen*, later *parfourmen*, wich is itself from the old French *parfournir* – *par* ('thoroughly') plus *fournir* ('to furnish') – hence *performance* does not necessarily have the structuralist implication of manifesting *form*, but rather the processual sense of 'bringing to completion' or 'accomplishing'. To *perform* is thus to complete a more or less involved process rather than to do a single deed or act. To perform ethnography, then, is to bring the data home to us in their fullness, in the plenitude of their action-meaning.

In this sense, it is also possible to define performance as a creative process by which the researcher experiences the mental categories of the studied people. This characteristic is absolutely central: in this case the creative aspect does not develop by applying a set of rules, but unfolds through a sort of *crossing-over* in which interaction is always a relationship that starts something new (knowledge in this case). According to Victor Turner (ibid., p. 79) the idea of performance takes the form of a process and not of 'rules or rubrics'. Rules here are mainly functional to the framing of the process, because 'a river needs banks or it will be a dangerous flood, but banks without a river epitomize aridity.'

In this context, a researcher's creativity is expressed in a different way, through a process that involves the deepest aspects of personal and social life. He follows the Latin meaning of the term 'creativity', as the creation of something that did not exist before (new mental categories). This is possible through a process that does not follow steps to control the hypothesis, but that proceeds in an experiential level. The frame by which the researcher interprets actions and behaviours changes. He begins accessing interpretations of reality that he could not think of earlier. In this case, he refers to what Chomsky (1964) defines as *rule-changing creativity*. The *performance process* is the one which leads the researcher to call into question his way of interpreting a particular portion of reality, and to change his interpretation when achieving the comprehension of the mental categories of subjects with whom he gets in touch. Through rule-changing creativity it is the conceptual frame that changes, which modifies itself, and allows us to reposition situations, thoughts, and interpretations of the social reality.

From another point of view, in social sciences, performance processes may entail different levels of interaction and involvement. This implies building trust with the people the researcher interacts with, to understand what they expect from him, as well as what role people recognize him playing. To face those issues means to manage unexpected situations, and this is never an issue of application of rules, but is a relationship, a crossing-over.

The image and behaviour of the knowing subject certainly reflect the categorization he uses to learn about the world. Such categories will be perceived and will influence the subjects studied, regardless of whether the study entails a relationship with an interviewee, an interaction with a participant in a focus group, or a sharing of life with a group of people during ethnographic research. For example, commenting on ethnography, Mario Cardano (2003, p. 126) emphasises that:

> Fieldwork starts with a particular rite of *status* inversion: the observer becomes the observed object by the 'natives' who, from the few evidences offered during initial encounters, legitimately try to understand whether, and to what extent, they can trust him. (author's translation)

A PRIORI CATEGORIES OF UNCONSCIOUS LEVEL

It would be misleading to think that the research activity will be devoid of problems at an heuristic level just by using a low level of structuring. Even in the extreme case in which the researcher could observe whatever aspect of reality, he cannot avoid – in the act of orienting his sight – scanning reality through his mental categories – perhaps unconsciously – and filtering information collection. He cannot rid himself of what he is and has been, thus bringing with him a reference framework to interpret and recognize reality at an unconscious level. According to Alfred Schutz (1932; Muzzetto, 1997) this means moving in a pre-logic level of consciousness, in which one brings back objects to known 'typification'. This process is placed into a deep level of conscious that works on automatisms. This leaves space for the possibility of turning to categories that the researcher already has, but which are not adequate to understand a new context. Still, in the early stages of research this seems to be unavoidable, for example when 'we identify a specific [empirical] referent as the member of a class . . . that we have already constituted (for example a table/a bicycle/a human face/a woman's voice/a rock song/a nervous disorder)' (Marradi, 2001, p. 13, author's translation). The unconscious process by which one assigns a referent to a class is called by Aristotle (*Posterior Analytics*, II.19) 'intuitive induction'. Cognitive psychologists call it categorization, whereas Schutz (1932), remarking on an ordering function of perceptive chaos that allows social interaction, uses the term 'typification' (1932, 1959; cf. also Husserl, 1939).

Nevertheless, it is only with late post-positivism that observation of reality has come to be no longer considered a cognitive activity that leads to a pure and incontrovertible awareness. The observation process is influenced by the researcher's cultural context, by perception and knowledge about psychological mechanisms that lead actions and, in the end, by the modalities of the observation. Thus, observation is an activity that collects information through the knowing subject. He will interpret the reality through his categories that can be cultural, learned by means of socialization, or learned by belonging to researcher's scientific community. The first two can vary across individuals and inside the individual as well. Although a scientific community has its own cultural and socialization dimensions, there are aspects of the specific categories of science which frame the world following more rigorous procedures: they are more stable, less ambiguous, capable of allowing a satisfying shared observation. For these reasons, observed data stop providing that certitude which has been a central linchpin in positivist thought. Rather, the reflection on how theory penetrates in observation processes becomes central and underpins the thesis of theory ladenness[8] (Hanson 1958; Brewer and Lambert 2001).

Hanson (1958) undertakes a philosophical reflection about epistemological problems in the field of particle physics. He concludes that to perceive even an ordinary object of everyday life, it is not enough to see it. Rather, human beings need to have a preliminary knowledge about it. Likewise, for an individual to obtain information from perception, the objects encountered must be identified. To do so, the individual needs access to a relevant set of information. Hanson gives some examples of objects that contemporary people directly perceive because they are familiar to them, but that people of past generations would not recognize, being extraneous to them. Hanson (1969) suggests that in seeing there is more than being blindingly obvious. What an individual sees is what his consciousness and education allow him to see, and what he learns about what he observes is not determined only by the object that he is looking at, but mainly by what he already knows.

Affirming that to hold from the beginning the whole set of categories which are needed for the correct interpretation of a new context would mean having an omniscient mind (Hayek, 1945). Moreover, it would mean deceiving ourselves with the belief of being able to observe free from any a priori category – even unconscious – which is equally absurd.

In *Critique of Pure Reason*, Immanuel Kant (1781) argues that reason sees only what it produces by itself: space, time and categories intervene in the phenomenon formation and the empirical data are constructed transcendentally through the a priori functions of sensitivity and intellect. Also, following Simmel (1900, 1950) it is possible to consider a priori comprehension based on pre-formed categories. However, because it is referred to as social reality, his idea of a priori does not seem to respect the fixity which is typical of the Kantian conception. The a priori by which the people recognize reality are modifiable by experience, they depend from a space–temporal dimension, they can vary from individual to individual and also inside the individual himself (Boudon, 1989; Migliozzi, 1996). This is a central process in that it allows the evolution of mental categories through experience. Such a process implies that to establish a contact with the knowledge shared by a group of people presupposes the willingness to open one's self to learning, and a capacity to modify one's thought categories in evolutionary terms.

THE CASE OF THE MUTOIDS' ITALIAN COMMUNITY

In 2006 I conducted ethnographic research in the Mutoids community based in Santarcangelo di Romagna, Italy. For six months I lived with them and shared the everyday life of this group of people. Mutoids are a small community composed of about 20 people who have been based in

Source: Author's photographs.

Figure 12.2 Photographs from The Camp. Top, the central monument, a sort of post-bomb dolmen; and bottom, a Mutoid sculpture

an ex-gravel pit in the suburbs of Santarcangelo, near Rimini, since 1990. The initial core of the community was created by people of English and Scottish origins, but other people from Italy joined the initial group later on. They live in caravans or old buses that have been transformed into houses. 'The Camp' is about 100 square metres of unpaved space where houses are disposed, lorries and cars are parked, and where wrecks that the group recycle are piled up. Members of the Mutoids live principally thanks to their recycling art, 'mutating' wrecks and other goods discarded by society into sculptures, mutant machines, installations, or doing performances inspired by a post-nuclear future. In The Camp's main square a great monument rises up. It was constructed by the Mutoids when they arrived in the gravel pit and is made of two lorries knocked vertically in the ground. They are linked together like a dolmen, but in a contemporary – post bombardment – version. Other sculptures are dispersed across The Camp and in front of the houses of their creators.

The only way to study and understand a way of life which immediately seems to be so different from the mainstream one to which I belong, was to apply a qualitative research design. A questionnaire would not have been an effective tool to collect data about them. This is because of the difficulty and perhaps uselessness of writing questions about a context that is radically unknown, following an *etic* cognitive perspective. The categorial *iato* (hiatus) that separated me from a comprehension of their existence could be filled only with a direct experience of their life.

For example, in the first period I spent in the field I did not really understand the role of the wrecks and other things jumbled in The Camp. I just did not wonder anything about them, I took them for granted, like a naïf frame populated by this different people. But this was a central issue that I did not recognize immediately. I was too much concentrated on gaining their trust and on the avoidance of doing something wrong, to pay attention to what was abundant in The Camp and which my mind recognized automatically as rubbish. Rubbish: probably few things could have been more misleading. To illustrate, when Connie, one of the Mutoids, invited me to her birthday party at The Camp, after having celebrated all night, bottles, plastic dishes and exploded balloons were jumbled all over the ground of the principal square. I had to go home to sleep a couple of hours, but I promised myself to be back in the morning to help with the cleaning up. The morning after, when I arrived at The Camp, I was really surprised to find that everything was neat and ordered. There were no signs pointing to the party of the night before: no bottles, no plastic dishes, not even a little piece of waste paper lying on the ground. I found that circumstance very strange. Why should people who live surrounded by rubbish clean so fast and meticulously after a party? Something did not match with my pre-existing habits of thinking and behaving. I could have ignored that contradiction, but something, possibly intuition, suggested to me that there was some relevant element I did not guess.

The only way to give sense to the whole situation was to state an hypothesis that (if true) explained as normal the event that I could not understand at my level of comprehension. This is an abductive[9] reasoning. Because of its logic construction, this is the only inference that allows us to introduce new knowledge in reasoning, to be more precise, that lead us to a conclusion that is not included in the premises.

That was the first time I really understood the function of the wrecks jumbled all over The Camp: It was not 'rubbish' but the discards of society that the Mutoids pile up and then use as the prime material for their houses, sculptures and performances, or for their everyday life.

Moreover, they are extremely sensitive to environmental issues: inside the community they practise the separate collection of rubbish that they obviously produce and discard, and outside they collect things that other people consider as rubbish, but they use as their necessities. I could not realize that without experiencing their reality. Furthermore, I could not predispose a tool to collect data about something I could not previously describe. The role of creativity, in this small example, was to change the rule through which I looked at the Mutoids' reality during the period I stayed in the field, and to face every emerging task due to the relationship with them.

CONCLUSIONS

Epistemological and methodological frameworks are central in a discussion of knowledge production. They contribute to understanding the use of creativity along the distinction I have borrowed from Chomsky (1964). Surveys constructed to control hypotheses are oriented by the knowledge codified by the scientific community. This initial predisposition appears in a more subtle way even when the researcher uses a qualitative approach. In this chapter I do not argue that qualitative approaches are superior to quantitative ones. Rather I critically analyse some differences between them; differences that are often well used in order to best exploit the potential of both of these approaches by a triangulation of different techniques.[10]

A discriminating factor in choosing an approach is based on the level of the existing knowledge pertaining to a particular phenomenon. If the researcher can describe that portion of reality in a satisfying manner, he could consider his knowledge as sufficient and set a structured tool for information collection. On the contrary, when he does not have – for any reasons – a level of knowledge that allows him to describe a specific portion of reality, he cannot use a structured tool, but he first needs to experience that reality.

The quantitative approach is characterized by *rule-governed creativity*, the qualitative one by *rule-changing creativity*: these are two models of creativity that the Chomskian vision links to a set of rules. Thus in the first case the creation of the tools by which the researcher collects information is submitted to a set of rules related to substantial and procedural competences. In the second case the creative phase does not have a place in the creation of a tool, but rather in a performance, that is to say in the modalities of interrelationship between individuals, through which the researcher comprehends the mental categories that people use to give sense to their reality.

Another difference about creativity regards the phases of knowledge production during research design. In the quantitative approach the creative phase is exhausted before going to the field by a unilateral act of the researcher. In the qualitative approach the creative phase is during the period in the field, mainly because it needs a continuing dialogic effort between the researcher and the subjects involved. The characterization of research phases according to their degree of creativity puts emphasis on the superior flexibility of the qualitative approach that in the field allows – by means of experience – for the consideration of unforeseen cognitive ways out of the researcher's categories.

NOTES

1. This argument is widely debated: cf. Reichenbach's classic distinction between context of discovery and context of justification (Reichenbach, 1951). For a perspective on qualitative approach, cf. Schwartz and Jacobs (1979); Linstead (1994); Maso (1995); Kelle (1995). For the quantitative approach, cf. Coombs (1964); Corbetta (1999); Bruschi (2005); Marradi (2007).
2. In what follows, for stylistic reasons and to facilitate the reading, I shall omit the female pronoun.
3. The emic/etic dyad was born and developed in the linguistic field and afterwards enriched in the cultural anthropology context. The first conceptualization is due to the linguist Kenneth L. Pike, who started the debate on the twin category from 1954 (see also Headland et al. 1990). In this context, to follow an *etic* perspective means to adopt the observer's point of view using previous knowledge of the reference group to which one belongs. Instead, an *emic* approach tries to achieve comprehension through the category of the examined culture.
4. By 'tool' I mean the medium between the researcher and reality which the researcher uses to collect information.
5. Here I am referring specifically to research that follows the hypothetical deductive model in the quantitative approach; in this approach, models differing from the

hypothetical deductive one can be found, such as factor analysis or principal component analysis, which I do not consider here. For those topics, cf. Bruschi (1999), Corbetta (1999).

6. Sensitivity is a property of the modalities through which a variable is codified, e.g. the same variable could be operationally defined by three response modalities (less sensitive) or by five (more sensitive). See also Marradi (1993, pp. 35–6).
7. For the issue of simulation in social science, cf. Moretti (1999, 2005).
8. 'Theory-ladenness' means loaded with theory. The expression 'theory laden' refers to observations and perceptions.
9. For a perspective on abduction see Peirce (1960); Eco and Sebeok (1988).
10. In the social sciences, the term 'triangulation' is usually adopted to mean the use of more than one technique, coordinated inside the same research project, to collect information about the same phenomenon.

BIBLIOGRAPHY

Amabile, Teresa M. (2001), 'Beyond talent: John Irving and the passionate craft of creativity', *American Psychologist*, **56**(4) April, 333–6.

Arlow, Jacob A. (1982), 'Scientific cosmogony, mythology, and immortality', *Psychoanalytic Quarterly*, **51**, 177–95.

Bickman, Leonard (1976), 'Observational methods', in Claire Selltiz, Lawrence S. Wrightsman and Stuart W. Cook (eds), *Research Methods in Social Relations*, 3rd edn, New York: Holt, Rinehart & Winston, pp. 251–90.

Blumer, Herbert (1969), *Symbolic Interactonism: Perspective and Method*, Englewood Cliffs, NJ: Prentice-Hall.

Bonnefoy, Yves (ed.) (1992), *Greek and Egyptian Mythologies*, Chicago, IL: University of Chicago Press.

Boudon, R. (1989), 'La teoria della Conoscenza nella "Filosofia del Denaro" di Simmel', *Rassegna Italiana di Sociologia*, (4), 473–500.

Brewer, W.F. and B.L. Lambert (2001), 'The theory-ladenness of observation and the theory-ladenness of the rest of the scientific process', *Philosophy of Science*, **68** (3), S176–S186.

Bruschi, Alessandro (1999), *Metodologia delle scienze sociali*, Milan: Bruno Mondatori.

Bruschi, Alessandro (2005), *Metodologia della ricerca sociale*, Rome and Bari, Italy: Laterza.

Cardano, Mario (2003), *Tecniche di ricerca qualitativa*, Rome: Carocci.

Chomsky, Noam (1964), *Current Issues in Linguistic Theory*, 6th edn, published 1975, Paris: Mouton.

Coombs, C.H. (1964), *A Theory of Data*, Ann Arbor, MI: Mathesis.

Corbetta, Piergiorgio (1999), *Metodologia e tecniche della ricerca sociale,* Bologna: Il Mulino.

Cowan, R., P.A. David and D. Foray (2000), 'The explicit economics of knowledge codification tacitness', in *Industrial and Corporate Change*, **9**(2), 211–53.

Cozzi, Donatella and Daniele Nigris (2003), *Gesti di cura. Elementi di metodologia della ricerca etnografica e di analisi socioantropologica per il nursing*, Milan, Italy: Colibrì.

Eco, Umberto and Thomas A. Sebeok (eds) (1988), *The Sign of Three. Dupin, Holmes, Peirce*, Bloomington, Indianapolis: Indiana University Press.

Gobo, G. and A. Tota (1994), 'Creatività e riflessività. Memorie di un gruppo di ricerca', in A. Melucci (ed.), *Creatività: Miti, Discorsi, Processi*, Milan: Feltrinelli.
Hanson, Norwood Russel (1958), *Patterns of Discovery: An Inquiry into the Conceptual Foundations of Science*, Cambridge University Press, Italian translation (1978), *I modelli della scoperta scientifica. Ricerca sui fondamenti concettuali della scienza*, Milan: Feltrinelli.
Hanson, Norwood Russel (1969), *Perception and Discovery*, San Francisco, CA: Freeman-Cooper.
Hayek, F.A. (1945), 'The use of knowledge in society', *American Economic Review*, **35** (4), September, pp. 519–30.
Headland, T.N., K.L. Pike and M. Harris (eds) (1990), *Emics and Etics. The Insider/Outsider Debate*, London: Sage.
Husserl, Edmund (1939), *Erfahrung und Urteil* [*Experience and Judgement*], Prague: Akademia Verlagsbuchhandlung, English translation (1973) by James S. Churchill and Karl Ameriks, Evanston, IL: Northwestern University Press.
Kant, Immanuel (1781), *Kritik der reinen Vernunft*, Königsberg, Germany: Reiger.
Kelle, U. (1995), 'Theories as heuristic tools in qualitative research', in I. Maso, P.A. Atkinson, S. Delamont and J.C. Verhoeven (eds), *Openness in Research*, Assen, The Netherlands: Van Gorcum.
Linstead, S. (1994), 'Objectivity, reflexivity and fiction: humanity, inhumanity and the science of the social', *Human Relations* **XLVII**(11), 1321–45.
Marradi, Alberto (1993), *L'analisi Monovariata*, Milan, Italy: FrancoAngeli.
Marradi, Alberto (1997), 'Esperimento, associazione, insieme non standard', in Gianfranco Bettin (ed.) *Politica e società. Saggi in onore di Luciano Cavalli*, Padua: Cedam, pp. 675–92.
Marradi, Alberto (2001), *Sai Dire che Cos'è una Sedia?*, Rome: Bonanno.
Marradi, Alberto (2007), *Metodologia delle scienze sociali*, Bologna: Mulino.
Maso, I. (1995), 'Trifurcate openness', in I. Maso, P.A. Atkinson, S. Delamont and J.C. Verhoeven (eds), *Openness in Research*, Assen, The Netherlands: Van Gorcum.
Migliozzi, Daniela (1996), 'Le interazioni della vita quotidiana nella sociologia formale di Georg Simmel', in G. Guarnueri and E. Morandi (eds), *La Metodologia nei Classici della Sociologia*, Milan: FrancoAngeli.
Montesperelli, Paolo (1998), *L'intervista ermeneutica*, Milan, Italy: FrancoAngeli.
Moretti, Sabrina (1999), *Processi Sociali Virtuali*, Milan, Italy: FrancoAngeli.
Moretti, Sabrina (2005), *Modelli e Conoscenza Scientifica: Problemi di Formalizzazione nella Ricerca Sociologica*, Milan, Italy: Guerrini.
Muzzetto, L. (1997), *Fenomenologia, Etnometodologia: Percorsi della Teoria dell'Azione*, Milan, Italy: FrancoAngeli.
Nigris, Daniele (2003), *Standard e non-standard nella ricerca sociale*, Milan, Italy: FrancoAngeli.
Peirce, C.S. (1960), 'A neglected argument for the reality of God', in C. Hartshorne and P. Weiss (eds), *Collected Papers of Charles Sanders Peirce, (Vol. 6: Scientific Metaphysics)*, Cambridge, MA: Harvard University Press. (originally published 1908).
Piore, M. and C.F. Sabel (1984), *The Second Industrial Divide*, New York: Basic Books.
Polanyi, Karl (1958), *Personal Knowledge: Towards a Post-Critical Philosophy*, Chicago, IL: Chicago University Press.

Reichenbach, H. (1938), *Experience and Prediction*, Chicago, IL: Chicago University Press.

Reichenbach, H. (1951), *The Rise of Scientific Philosophy*, Berkeley, CA: University of California Press.

Schutz, A. (1932) Italian translation, (1974), *La Fenomenologia del Mondo Sociale*, Bologna, Italy: Il Mulino. 'Der sinnhafle Aufhau der sozialen WeH: Eine Einleitung in die verstehenden Soziologie, Vienna: Springer, republished 1960 and in 1974, Frankfurt: Suhrkamp English translation by G. Walsh and F. Lehnert (1967), *The Phenomenology of the Social World*, Evanston, IL: Northwestern University Press.

Simmel, Georg (1900), Philosophie des Geldes, Leipzig: Duncker of Humblot, Italian translation (1984), *Filosofia del denaro*, Torino, Italy: UTET (1984).

Simmel, Georg (1950), *The Sociology of Georg Simmel*, translated and edited by Kurt Wolff, reissued in 1964, New York: Free Press.

Simmel, Georg (1976), *Il conflitto della cultura moderna e altri saggi*, Rome: Bulzoni.

Strauss, A.L. and J. Corbin (1990), *Basics for Qualitative Research: Grounded Theory Procedures and Techniques*, Newbury Park, CA: Sage.

Schwartz, H. and J. Jacobs (1979), *Qualitative Sociology. A Method to the Madness*, New York: Free Press, Italian translation (1987), *Sociologia Qualitativa. Un Metodo nella Follia*, Bologna: Il Mulino.

Turner, Victor (1982), *From Ritual to Theatre. The Human Seriousness of Play*, 2nd edn, 1992, New York: Paj.

PART V

Conclusions

13. Positioning order, disorder and creativity in research choices on local development

Silvia Sacchetti and Roger Sugden

'fatti non foste a viver come bruti ma per seguir virtute e canoscenza'
[*You were not made to live as brutes but to pursue virtue and knowledge*]
Dante Alighieri, 'Inferno', *Divine Comedy*, Canto XXVI

1. PLACES IN ORDERED AND HIERARCHICAL SYSTEMS

What the contributions in this volume have tackled is the possibility for an institutional dynamism that impacts not only on the characteristics of 'localities' and their position in a hierarchical and ordered system of relationships, but on the nature of the system itself. Is hierarchy a necessary and desirable condition for the development of economies? This is a perspective that, in line with Veblen and Hodgson's interpretation, addresses phylogenetic change (Veblen, 1898; Hodgson, 1993).

The idea of hierarchy may be positioned in some philosophical perspectives and views of the world which have associated the notion of hierarchy with that of place. In the *Divine Comedy*, written in the 14th century, the author, Dante Alighieri, undertakes a symbolic journey in the afterlife. His course, as defined by the Christian tradition, is articulated through the places and the gatherings which delineate an imaginative space, whose structure reflects the Greek and the Jewish–Christian cosmology (Stabile, 2007). For the latter, heavens are structured in different layers, each one of which – the empyrean or the place of peace where the chosen sit; the stars and the planets; the living beings, who are born and perish within the terraqueous globe – is subject to the natural law of circular and uniform motion or, in other words, each layer is bound to perfection. In 14th century Italy, when Dante writes – localisms and egoisms amongst and within cities were associated with precariousness, disorder and uncertainty. It was a time of conflicts during which hierarchy was advocated as the ordering and unifying principle, not only at the level of politics, but also for linguistic reasons (ibid.).

The neo-Platonic hierarchy of heavens, as Stabile (2007) suggests, defines space by means of subsequent specifications, moving from an indistinct totality to a multiplicity of elements, each with its own characteristics. The notion of *place* substitutes that of indistinct *space*. Each place is situated, that is, it is defined by what surrounds it, by the position it occupies in the world hierarchy. Because of the concentric structure which results from this notion, places find their own identity within an *ordered and hierarchical* system, so that each place's coordinates are defined indeed by other places (ibid.). In the last resort, each place – whether sited at a lower or upper layer – contributes to define at least another place, thus conferring significance to the system.[1]

2. ECONOMICS, ORDER AND HIERARCHY

Economics probably epitomises the idea of an ordered system in which change is normally contemplated and explained by identifying dynamic elements within pre-defined paths of development. If we consider the idea of a tendency towards 'normality', which was criticised by Veblen ([1898] 1998) and which the neo-platonic perspective on space emphasises, we can perhaps appreciate even more why in so doing economics 'can not be an evolutionary science' (ibid.). Rather, it tends towards an exogenous 'topographic order', within which each element (e.g. resources, regions, but also values and preferences) has its own place. Such an order obeys a hierarchical principle at the micro, for instance the exogenous preferences assumption, as well as at the macro level, for instance in the debated and still influential theory of development stages (Rostow, 1959), for which localities follow an obliged path towards 'normality' sketched, for example, in sectoral terms, or with respect to wealth, as well as through comparative analogies with the competitiveness of other places. This perspective delineates a static and marked lane towards predefined objectives, assuming an order of preferences at the macro level as well.

More generally, Veblen's critique to the idea of equilibrium as the law defining a tendency to 'normality' can be traced back to the neo-platonic ordering principle. Consistently, resources return to the place that an efficiency-maximising system confers to them. As argued by the Austrian school, the market and the price mechanism would serve this purpose (Hayek, 1945). However, as transaction cost theory has suggested, price signals might not be sufficient to disentangle the complexity attached to economic decisions. The knowledge required for choosing amongst alternatives and the associated costs would require different forms of economic governance, according to the classic argument that opposes markets and hierarchies. In that context, hierarchy serves the purpose of creating certainties in a system that would otherwise be characterised by a great deal of uncertainty.

Within the usual hierarchical organisation, such as the large corporation described by Chandler (1977), authors have read the introduction of the managerial function as a means for 'buffering' the environmental uncertainty and complexity of expanding markets (Langlois, 2003). This, it is argued, was needed to create a more stable context in which investment commitments to capital-intensive sectors could be implemented, and production could be run at its minimum efficient scale. That analysis is consistent with Knight (1921). He contended that, without uncertainty, management would be concerned with the coordination of activities in a routine sort of manner and without responsibility.

However for Knight (1921, p. 268; quoted in Coase, 1937, p. 400; emphasis added):

> With the introduction of uncertainty – the fact of ignorance and the necessity of acting upon opinion rather than knowledge – into this Eden-like situation, its character is completely changed. . . . With uncertainty absent man's energies are devoted altogether to do things; . . . in a world so built that perfect knowledge was theoretically possible, it seems likely that all organic re-adjustments would become mechanical, all organisms automata. With uncertainty present, doing things, the actual execution of activity, becomes in a real sense a secondary part of life; *the primary problem or function is deciding what to do and how to do it.*

As a consequence of uncertainty, Knight perceives it as 'imperative' to have direction (or, in his words, 'cephalization') as a form of governance – a point that Coase (1937) addresses and criticises by arguing for the role of the price mechanism under particular circumstances.

The for-profit firm, the prevailing economic organisation of capitalist systems, is a typical hierarchical structure. Its governance does not have democratic roots (Borzaga and Tortia, 2009) or, to put it differently, does not originate from a delegation process. As a matter of fact, whilst with democracy we say that the people are sovereign (although it is through a delegation process that representatives get the power to take decisions collectively), in the for-profit firm the command and decision-making power belongs to those who retain property rights and alternatively to the management or, more generally, to a restricted group of people (Cowling and Sugden, 1998). In that case, we cannot talk about a delegation process which springs out of the stakeholders' will (workers, for instance, or the local community as a whole) and which is based on a democratic rule (compare delegation with the contractarian perspective put forward by Sacconi and Degli Antoni). The founding institution for such a firm is property rights which allow for strategic control over production decisions. Consistently, by means of a labour contract, the employee hollows out to the employer the power to decide over the use of his/her time in exchange for a salary (cf. Screpanti, 2001). Moreover, beyond property rights, the firm can eventually exert its control by means of market power over consumers, suppliers and subcontractors (Cowling and Sugden, 1998).

There is, for these reasons, a substantial difference compared to, for instance, democratic systems. Under current forms of democracy, the governed – in principle – are also those who govern, by delegation. Consistently, state institutions – with that system of counterweights amongst powers initially advocated by Montesquieu ([1748] 1989) – should be functional to an ordering principle that serves the public (or publics, according to Dewey, 1927). That is to say, to recall a Smithian argument, that institutions should serve the purpose of safeguarding the positive and negative freedom (Berlin, 1969) of all actors, in particular those with the weakest (economic) power.

To state it differently, institutions may also be seen as instrumental to the need of not falling back to that Hobbesian *state of nature* where man is wolf for other men.[2] We are in front of a very different idea from that of 'free market competition' (across individuals and, at a meso level, across firms) taken as the ordering principle of economies. That notion of competition put forward by Alchian (1950) as well as by Hayek (1982) has heavily shaped the understanding of economic development in terms of the 'survival of the fittest'.[3] As Hodgson (1993) has noticed, the fittest organisations have been identified with the prevailing organisational structure of contemporary economies, the for-profit firm. However even the simple fact of representing the majority of firms, does not mean that they are the most efficient, or the better suited organisational structures for societies. Furthermore, the liberalist approach to economics focused on maximisation has, paradoxically, eliminated free will (ibid.). In fact, once we introduce uncertainty, orthodox rational choice stops being founded on the purposeful reasoning in individuals, whilst explaining action in terms of mere 'programmed responses to the circumstances in which those agents are placed' (Loasby, 1991, p. 1, quoted in Hodgson 1993, p. 218).

Following Knight (1921), under uncertainty production requires exclusive decisional structures (the managerial function) which introduce power asymmetries within but also amongst organisations. But because of the relative indeterminateness associated with intentionality of action (as against the maximisation assumption on rational agents) and of the imbalances created by uncertainty, the market – although assimilated by many to the idea of freedom – without rules would probably be a good approximation of the state of nature and precariousness, of the uncertainty and disorder that the neo-platonic philosophy wants to avoid with its idea of cosmology.

3. THE HUMAN PERCORSO TOWARDS SELF- AND DE-LIBERATION

Under uncertainty, whether the governance mechanism is exclusive or participatory, we do not find such a thing as complete information (Coase, 1937; Williamson, 1985) and knowledge (Hodgson, 2005), where the notion of knowledge is dynamic and related to the differentials in learning and forgetting of human beings, organisations and society. To get round these limitations, it has been argued that institutions play the role of a collective mind by 'remembering' in the place of individuals, organisations and society (Douglas, 1987), and by simplifying individuals' tasks (Hayek, 1967; Egidi, 1987).[4] That happens as habits and routines[5] indirectly replicate themselves by eventually activating specific behaviours.[6]

From the limited nature of human knowledge follows its perfectibility. We can, in other words, expect knowledge to evolve and impact on humans' understanding of their own desires, preferences and choices both as individuals and as publics. Institutions, such as habits and routines, indeed provide an ordering principle as they can be conceived as shared dispositions or propensities (Dewey, 1922) on which individuals and organisations can activate higher deliberation processes (Hodgson, 2003).

It is institutions, for example, which can be thought as defining the space of production and exchange and make it work towards an order. Such an order, we suggest, would be in turn a *means* for achieving socially desired and democratically identified objectives, rather than an *end* in itself, the argument being that this specific conceptualisation of order allows uncertainty to be present. From this perspective, the rules that through democracy are shaped by publics should – in principle – avoid social and economic exclusion. Deliberative processes, by involving rather than excluding, would allow for differences to emerge across actors and places, thus encouraging and offering a more favourable context to the flourishing of variety.[7] This is not a straightforward process, as it requires the creation and use of *deliberation procedures* for mutual learning and consensus formation, including horizontal communication amongst a plurality of publics, as noticed by Sugden, Te Velde and Wilson (Chapter 10).

At its heart, we could say that deliberation constantly applies the principle of the *Civitas*.[8] *Systematically*, according to what has been argued in

Chapter 9, deliberation becomes the attitude reflecting the need to include those publics which are, at different times, interested in specific decisions. With respect to the neo-platonic and Dante's-like order, there is a change in the coordination mechanism. Under a prospect of inclusion, governance mechanisms do not reflect hierarchical principles through, for example, the government of a 'sovereign' – who might be enlightened but certainly not omniscient and not even the repository of justice alone, as Young (2002) noticed. Rather, we suggest that coordination occurs by means of the democratic and ordered deliberation amongst publics (Sacchetti and Sugden, 2008). The objective is to get to a shared decision by means of a deliberative discussion and communication, rather than through a decision imposed from the top of a hierarchy to which a contractual resolution has conferred the right to decide for others.

Deliberation, therefore, is conceived here as an ordering principle (ibid.) which, differently from other procedural criteria, is also inclusive and egalitarian (Young, 2002). On this, in particular, Adaman and Devine (2001) have argued that deliberation allows for an evolutionary process of collective learning. Hodgson (2005), in contrast, argues that innate cognitive limitations can restrict the scope for deliberative decision-making. Stemming from a recognition of the limitations intrinsic in the nature of human rationality as well as in knowledge and learning both as individual and collective processes (Rizzello, 1997; Bruni and Sugden, 2007), we suggest that, in order to understand its limits and potentials, an analysis of deliberation needs to recapture the institutional focus of Veblenian origin on habits, as well as the individual dimension of interpretation and choice.

What accomplishes deliberation before, during and after the communication and decisional process points towards the individual's cognitive or thinking abilities, i.e. to his/her ability to understand a specific phenomenon *eventually against existing social habits* (Dewey, [1910] 1991). This stage, however, is reached through specific behaviours which are mediated – at a procedural level – by those same deeply rooted social habits, including the impact of specific inner beliefs on behaviours that can lead to effective learning. That paradoxical situation is emphasised by Veblen (1898) but also by Dewey ([1910] 1991), where he reasons that in order to overcome the 'propagation of error' (ibid., p. 21), the mind must be trained (e.g. through education) to the formation of habits that privilege that kind of thought which can 'discriminate between beliefs that rest upon tested evidence' and those that conversely are based on 'mere assertions, guesses, and opinions' (ibid., pp. 27–28).

The link between one's dispositions or habits, their actual understanding and contextualisation, as well as the behaviours that follow is mediated, in the last instance, by the individual and his/her own self (Hayek, 1952; Polanyi, 1967). We, therefore, suggest that *de*-liberation[9] is an element of humans' curiosity, of the desire for knowledge which ultimately aspires to *self*-liberation (e.g. from 'false beliefs' as Dewey would say). The latter (which would be a privileged terrain for philosophy and psychology) is, from a complementary perspective, the field over which artistic expression moves and, more generally, where the *thinking individual*, under particular conditions, chooses and acts.

Motivational aspects and preferences, as stressed in different parts of this volume, are relevant aspects for an understanding of the '*percorso*' that each individual chooses to undertake. In that context, to give space to diversity represents a necessary recognition for the differences across individuals and – as Grönblom and Willner notice in Chapter 6, recalling Elster (1986) – for the multiplicity of selves that may coexist at a given point or unfold over time.

As an illustration, Bertolucci (Bachmann and Bertolucci, 1973), describes the act of filming as the creation of a learning relationship between the film-maker and things, spaces and people, in search of one's own *self-liberation*.[10] This process, which does not refer to the arts only but which in the arts finds perhaps a privileged context, requires that individuals have the space to think, see, imagine and to let in the unforeseen. Freedom, from this point of view, relates to the possibility to exert one's own original *idle curiosity* (Veblen, [1898] 1998). Collective processes of de-liberation interact over time with the sphere of self-liberation, without substituting it, as the private sphere and the public sphere of reflection are complementary, although the latter could not exist without the former.

This links consequently to the dynamic element of human research, which may be subject to different degrees of intentionality and intensity. A consideration of 'time' allows talking of a 'percorso'. The film-making experience assumes a specific meaning in that respect:

> I still seek the very specific way of representing how time passes – that particular psychological passage of time which gives a film its style. Perhaps it is a matter of 'percorso', of how a man moves through time, in the historical and in the practical, daily sense. (Bachmann and Bertolucci, 1973, pp. 6, 7)

The reason we would rather refer to a 'percorso' in the search of self- and de-liberation (the individual and the publics) is that we want to take distance from the notion of 'evolution' in economics put forward by Alchian (1950) and Hayek (1982). Such approaches are based on the idea promoted by earlier social evolutionists such as Spencer ([1876] 1969), who privileged the notion of human 'survival' and of competition for survival, whilst with a focus on the 'percorso' we want to recapture essentially the intellectual and practical experience of human 'life'. It is the 'percorso' partly designed by the choices of human beings towards their liberation, towards understanding, but also affected by existing institutions and by their inertia (Veblen, [1904] 1996), as well as by path dependence on previous events (Arthur, 1989).

There is teleology in what we suggest. What, for instance, the notion of 'mutual dependence' across economic actors entails can be defined as a benchmark, an ideal type objective (Sacchetti and Sugden, 2003).[11] The objective in itself, although identified, could be unachievable, but that does not make it less desirable. From this perspective we reposition our analysis from 'winners' in the competition for survival towards a focus on the process.[12] This is why, against current trends, we suggest not to judge only from the achieved, but to focus on the running path, on the choices as well as on tendencies[13] *towards* (or, more dramatically, *against*) a desired objective.

A focus on the 'percorso' pulls us out from the culture of *achievement* in which we are submerged (Bachmann and Fellini, 1980–81), and recaptures a realistic,[14] less mythical perspective by putting emphasis on uncertainty, and on the dialectic between order and disorder (see also Christensen, 2007).

4. DISCOVERING 'LAWS', DESIRES AND PREFERENCES THROUGH EXPERIENCE

What, then, is the role of rationality in such a context? As Sugden (1991) notices, according to Jevons' transposition of Humean theory, rationality is instrumental, i.e. an instrument to achieve the maximisation of utility (Jevons, [1875] 1905). Such an end is possible if we assume that preferences are complete and transitive and consistent with choices, which are understood in the simplified context of 'the ordinary business of life' or the 'lowest rank of feelings' (ibid., p. 91).

Stemming from the particular debate on which Bruni and Sugden (2007) comment, we argue that instrumentality is not adequate to explain all economic interaction or, more broadly, all social interaction. A first criticism of instrumentality is that, at the conceptual level, by identifying rational action with the maximization of utility, human choice involves only programmed responses to an exogenously given objective, thus removing the purposeful behaviour of actors (Hodgson, 1993). Second, under orthodox rational choice theory, uncertainty disappears from the context in which humans make their judgements and chose (ibid.). Contextually, Alchian (1950), subsequently criticised by Penrose (1952), located maximising behaviour in evolutionary reasoning, arguing that only maximisers are bound to survive. Hodgson (1993) effectively notices that, by denying free will, orthodox theory made 'the individual a prisoner, not simply of the social environment, but of his or her immanent and often invariable preference functions and beliefs' (ibid., p. 218).

How, then, is knowledge achieved by individuals? Having emphasised the importance of the 'percorso', we would put forward a notion of rationality based on experience (Dewey, [1910] 1991), including the use of reason, or critical thinking, as a fundamental component both at the individual and at the collective level.

A number of contributions have emphasised that people's mental capacities to learn from experience are subject to limitations and errors (Bruni and Sugden, 2007; Hodgson, 2005). However, experience is here a synthesis of the use of reason and evidence at the level of the self (or multiple selves) and across different individuals at different times. Rather than transcending space and time, conclusions stemming out of this process would be contingent on the individuals and the publics involved in that particular deliberation process at a given time, as the identification of different places, individuals and publics would require further deliberation.[15]

The Deweyan notion of experience, in particular, is a synthesis of the personal and public dimensions in their interaction with the environment:

> Experience is the result, the sign, and the reward of that interaction of organism and environment which, when it is carried to the full, is a transformation of interaction into participation and communication. Since sense-organs with their connected motor apparatus are the means of this participation, any and every derogation of them, whether practical or theoretical, is at once effect and cause of a narrowed and dulled life-experience . . . What is distinctive in man makes it possible for him to sink below the level of the beasts. It also makes it possible for him to carry to new and unprecedented heights that unity of sense and impulse, of brain and eye and ear, that is exemplified in animal life, saturating it with the *conscious meanings* derived from communication and deliberative expression. (Dewey, [1934] 2005, p. 23, emphasis added)

Along the 'percorso', values, but also latent habits as well as objectives and preferences need to be constantly assessed or discovered. In this context, therefore, rational actions are those that lead individuals to continuously scrutinise and update their understanding of *actual* objectives, preferences[16] and desires,[17] consistently with their inner values, which are subject to a *process* of discovery as well. At the same time, through deliberation individuals relate their evolving laws, desires, objectives and preferences to those of other people and publics.

From that perspective, rational behaviour is therefore a 'meta-behaviour', as it leads to an appreciation of the 'percorso' and the dynamics in humans' search of values, habits, desires, objectives and preferences. Action (including imagination and thought) is needed at time 'T1' in order to explore inner wishes, learn and eventually reshape objectives and preferences accordingly at 'T2' before you can act at time 'T3', when further meaning is attached to events and thoughts, feeding them back as new inputs in the process of meaning and learning. Such a process is effected by the *opportunities and constraints* attached to the environment, to habits as well as to the human mind, but remains fundamentally a 'percorso' towards 'liberation', towards how to improve our understanding of who we are in terms of what we think acceptable principles are, following life experiences. This process builds upon itself and is based on experiencing things both at the level of the self and at the level of other publics. Such 'eagerness for experience, for new and varied contacts, is found where *wonder* is found' (Dewey, [1910] 1991, p. 31, emphases added).

Action, according to what we suggest here, is not to maximise pleasure or to minimise pain, as hedonistic approaches suggest. Nor it is a direct consequence of our tastes, as Pareto argued, because we might not know them. Rather, action is to improve our knowledge. This is possible only if action becomes 'experience' as the latter includes the thinking functions of the mind (Dewey, [1910] 1991).[18] The Cartesian '*cogito ergo sum*' (I think therefore I am) could be rephrased and extended here in terms of: *I experience therefore I am*. The Shakespearean question of being, from this particular perspective, could be interpreted as lining up the conditions under which an individual can proceed towards knowing him/herself and chose *consistently* with his/her perceived values, habits, rules and desires. The point about consistency implies the retrieval of instrumental rationality (or hypothetical imperatives, in Kant's terms), which however cannot exist without the initial 'wonder' or 'idle curiosity' which prompt in people the desire for knowing.

For human beings, knowing themselves and understanding themselves comes before any assumptions about maximisation of pleasure as being the objective or tastes. Knowledge may give pain, indeed. The question is not: how do human beings maximise pleasure, but how can they understand their values, their desires and create the opportunities to coherently discover preferences that are consistent with such (perceived) inner wishes and understanding of things. Critical thinking is essential to this process. *It is a way of acting and making sense of action at the same time.*

Particularly, we have argued (Chapter 9) that critical thinking can be stimulated by de-liberation and by activities such as the arts or science, which are actions impacting on our understanding and are, therefore, at the basis of a 'percorso' of desires and preferences discovery. We suggest that these develop when both the private and the public aspects are considered, that is, if preferences are sought whilst making sense of what is experienced *both as a personal reflection as well as interacting with publics.*

5. UNCERTAINTY IN LEARNING AND RESEARCH ACTIVITIES

In stressing the importance of intellectual curiosity and of the use of reason and evidence along the 'percorso', we would nonetheless highlight the significance of intuition. When discussing sense and sensibility, Docherty (2008) refers to the study of English and, more generally to 'the proper place . . . of literature . . . within a society'. He considers the deficiencies of having 'a form of knowledge that was not "lived", not actually "felt" at the inner level of sensibility' (ibid., p. 4). The result of such a form might be seen as 'a triumph of the industrialisation of the human spirit', a failure to appropriately balance 'sense or reason and sensibility or feeling' (ibid., pp. 4–5). The idea is that the study of literature might demand a balance, but we would argue that the same could be said for the study of economies.

One hypothesis is that an understanding of a poem, painting, machine or, indeed, economy that is not felt might be in some sense lacking. For example, an engineer or economist might first feel what is failing in a particular machine or economy and then make that feeling tangible, expressing it through reasoned argument and correcting the failure; but such reasoning and correction might not occur without the feeling, other than by chance. Another hypothesis is that sensibility is a determinant of people's capacity to be imaginative; that understanding of economies requires imagination; therefore that an academic studying an economy needs sensibility for the subject in order to unleash his or her full creative potential in exploring what is and might be happening in the economy. Without sensibility, imagination is constrained and understanding limited.

What does that appreciation of sensibility imply for academia? The idea here is that sensibility can guide individuals not only towards the knowledge of values, desires, preferences and objectives from an ethical perspective, but also from a scientific point of view. Our approach – in line, for instance, with what has been argued by Francesco Sacchetti in Chapter 12 – suggests to organise research and learning activities so as to nurture the intrinsic creativity of each participant. Chomsky (1975, p. 164) argues that the purpose of education is not 'to control' a person's 'growth to a specific, predetermined end, because any such end must be established by arbitrary authoritarian means; rather, the purpose of education must be to permit the growing principle of life to take its own individual course, and to facilitate this process by sympathy, encouragement, and challenge, and by developing a rich and differentiated context and environment'. Similarly, our concern is to forsake the attempted certainty associated with predetermined outcomes and instead to create discipline-spanning and discipline-fusing spaces that: (1) enable the positive freedom of scholars (faculty and students alike) to identify and pursue a unique academic agenda that both reflects and contributes to the development of each person's intrinsic creativity; (2) provide a sympathetic yet challenging environment that emphasises scholars' creative interaction, together and with other interested publics.

Rather than analysing predefined topics in search of predefined outcomes, the cost of which can be omission of potentially relevant and insightful issues, we suggest that topics and analysis be allowed to emerge by the creation of spaces where uncertainty and some degree of disorder are accepted, across disciplines, across methodologies, across publics, across social spheres, across localities. An objective is to be open to complementarities with others so that we might discover and pursue new and unique opportunities, while maintaining and developing core competencies as academics. This enabling of disorder and uncertainty is in interesting and stark contrast to what Khurana (2007) has argued to be a foundation of US business schools.

6. CREATIVITY AND UNCERTAINTY IN RESEARCH OVER LOCAL DEVELOPMENT STRATEGIES: A CASE FOR ACTION RESEARCH

The relationship between universities and territories has been shaped within policy and academic environments around the notion that academic research may benefit from case studies and observations based on regional experiences and, vice versa, that regions and localities may benefit from the insights that academia might give with respect to specific economic problems. This view stems, not least, from the way in which science develops, including, for example, the intrinsic interplay between deduction and induction, between the development of hypothesis and the observation of specific, perceived realities.

However, recent evolutions (think for instance about Porter's major international influence on policies for competitiveness and promotion of global chains (Porter, 2003)) show that economic thinking is heavily drawing on policy objectives and, not least, on the understanding that policy-makers have of the notions underpinning those objectives. One possible shortcoming for academics and economic theory is to lose analytical power and critical perspective or, in other words, to preclude the analysis of different ways forward, other than those currently embraced.

In May 2007 we organised a Festival on Creativity and Economic Development which was hosted in Gambettola, a 10000-inhabitant municipality in the Emilia-Romagna Region, in Italy. The initiative came from concerns coming out of our research interests as academics on the one hand and, on the other, it was the result of policy-makers' will and openness to find new ways forward. In particular, the idea was to stimulate and nurture people's own imagination and creativity, so that together they could shape and seize opportunities for critical thinking, for talking and discussing with each other, for respectfully sharing opinions and arguments, thereby increasing the diversity of ideas and perspectives, for finding mutually respectful and beneficial ways forward in social, economic and political activities.

This Festival included seminars and discussions involving leading academics and practitioners; these were sometimes essentially 'academic' and at other times targeted at wider publics. The seminars were all linked to a series of performances and exhibitions – sculpture, photography, dance, poetry, theatre, music, film – that explicitly focused on the relation between creativity and economic development. The presenters, performers and exhibitors came from various countries, and overall the one week event attracted some 3000 visitors.

At the outset, as academics, we had severe doubts that it was possible to encompass, in one event, academic analysis and artistic performances, whilst also actively engaging people from the local community. In this regard, however, the Festival combined the contributions of theatre actors, architects, citizens, economists, entrepreneurs, graphic artists, poets, policy-makers, political scientists, students.

To explore old and new issues in economic development in different ways meant to create new spaces as well as, borrowing some of the evolutionary terminology, 'new combinations'. Against the current approach to research, which relies on predefined issues and outcomes in order to increase certainty, the intention was to let the topics emerge by creating space across localities, across disciplines, across social spheres. Although increasing the uncertainty of outcomes, the new space and combinations would eventually dynamically change our routines as academics. And because routines involve a strong tacit dimension, they may not be easy to imitate (Dosi, 1994, p. 233), thus eventually opening up to new topics or to a different experience of interaction. Moreover, new routines have the potential for tooling up a university or a locality with distinctive capabilities, therefore providing a basis for differential performance.

The dangers associated with what we could call a holistic approach are that the creation of new spaces requires the acceptance of some specific risks, at least in the first stages of development. In particular, a major difficulty could come from habits of thought and routines, as academics might keep on sticking to what they know best. As the research environment changes, cognitive structures could fail to adapt quickly enough to recognise relevant objects and relationships. However, borrowing from industrial analysis, new opportunities can be developed, where 'such opportunities are partly a lagged function of "fomentation" "diversity" and "search activities"' (Dosi, ibid., p. 234). In this respect, academics might need to elaborate new routines, but they do not change their core competences as academics. Rather, they develop complementary assets which are crucial for discovering new opportunities.

A key feature of the Festival is that it was consciously *designed* to foster unforeseen outcomes, all being consistent with participants mutually respecting each others' experiences, skills, perspectives and aims. In seeking this, we deliberately introduced and sought a critical perspective on accepted approaches. In fact, we would argue that in practice the Festival did indeed yield unforeseen outcomes, in terms of analysis, understanding, and cooperation.[19] Moreover, it contributed to activate latent attitudes towards participatory governance by activating some radical changes in the local institutional structures governing the complex and dynamic set of publics related to the specific activities that characterise the town.

Experience with the Festival is also instructive on the issue of engagement with interested publics whilst retaining independence from those publics. In organising the event we had strong support from the Mayor of Gambettola on the basis that he and we would play explicitly different roles. Specifically, the Mayor's role was to take account of all political issues and to address them as political issues. He did not see our role as providing the political agenda and solutions. Rather, he drew from an independent position on the knowledge and stimulation provided by the event. Further uncertainties were associated with the location of the event, in a town that can be defined as peripheral to the nearby city. The realisation of the Festival proved that it was possible to conceive and deliver an event that would not normally be associated with the likes of peripheries, namely because of the centripetal forces affecting resource allocation across the institutional, productive and geographical hierarchy of places (Myrdal, 1957). A major element in allowing this could be identified in maintaining the integrity of the activity when the substance of it challenges the existing institutional order. In our specific case, an explicit choice needed to be made when faced with the suggestion of moving the event in the nearby more central town, possibly because hosting such events in peripheral places contrast with established habits of thought.

As for our self-perception of our own role as academics, we sought to set our agenda from an academic and not a political perspective; to show respect for the political agenda but without compromising our academic judgement; to have the final say on all aspects of the Festival, content and management. Building on the Festival experience, we suggest that careful consideration of the way in which academics relate with policy-makers (and therefore territories) might help to gain further insights on the nature and implications of such interaction. One of the current practices is the use of academia as a pool of knowledge that policy-makers can use (for example, in terms of consultancy) at convenient times to underpin strategic choices. However, we would argue that this sort of interaction, although useful in some contingencies, is restrictive of the broader role of academia on the one hand, but also limitative of the different forms that interaction between academia and localities may assume, on the other.

Reflecting on the experience, we then saw the opportunity of analysing and discussing the nature of academia and the implications of its relationship with different publics, both within and outside a university organisational border. In particular, coming from the Festival experience, we reflected on the meaning of academic freedom, stimulating inter-disciplinary discussion.[20] One major issue – on which we would not dare to formulate a tentative answer yet – is whether challenging existing practices and routines is, if not promoted, at least accepted when applied to academia itself. Still, this is perhaps not just applied research on creativity and economic development. It is about ourselves as academics, and how much we can push the limits of the paths and trajectories that have been already traced.

ACKNOWLEDGEMENTS

The authors would like to thank Ermanno Tortia and Romeo Casalini for having discussed various ideas presented in this chapter. Earlier work was presented at the Workshop on Ethnicity, Creativity and Economic Development, Åbo Akademi University, Finland, in December 2008; we are grateful to participants for discussion.

NOTES

1. Take for example the European Union White Book on Governance (COM, 2001), which identifies a global governance level with which the European Union central governments, regions, cities and the civil society relate.

2. Hobbes describes the birth of the Leviathan as follows: 'This is more than Consent, or Concord; it is a reall Unitie of them all, in one and the same Person, made by Covenant of every man with every man, in such manner, as if every man should say to every man, I Authorize and give up my Right of Governing my selfe, to this Man, or to this Assembly of men, on this condition, that thou give up thy Right to him, and Authorise all his Actions in like manner. This done, the Multitude so united in one Person, is called a Common-wealth, in latine Civitas. This is the Generation of that great Leviathan, or rather (to speake more reverently) of that Mortall God, to which wee owe under the Immortal God, our peace and defence . . . From this Institution of a Commonwealth are derived all the *Rights*, and *Facultyes* of him, or them, on whom the Soveraigne Power is conferred by the consent of the People assembled'. (Hobbes, [1651] 1996, Chapters XVII and XVIII, pp. 117–29).
3. See Hodgson (1993) for a detailed criticism and analysis.
4. According to Hayek (1967) institutions (such as language, law or money) emerge as the outcome of no particular will and not, as contractual (Hobbesian) theories suggest, through the rational design of individuals. From Hayek's evolutionary perspective, institutions are the result of evolution, through a process of adaptation and selection.
5. Routines, as distinct from habits, have been analysed in Hodgson (1993) using the psychological category of *procedural memory*. Procedural memory, differently from cognition or thought, does not 'have the capability of modelling the external world – that is, of storing representation of objects, events, and relations among them' (Schacter, 1990, p. 301, quoted in Hodgson, 2003, p. 375).
6. From an evolutionary economics perspective, Hodgson (2003) treats individuals' habits and organisational routines as the underlying causal elements (*genotype*, or gene in biology) that may generate specific behaviours (*phenotype*, or character in biology). We say 'may' because as Hodgson (ibid., p. 373) stresses by recalling the work of Veblen ([1898] 1998) and Dewey (1922), habits are capabilities, *propensities* which 'replicate indirectly by means of their behavioural expression'. Behaviours are therefore the manifest effect of (latent) habits. It is behaviour (the phenotype), in social systems, which is imitated and therefore replicated, thus allowing for *similar* habits to be acquired. Such prospect, which by grounding on the transmission of acquired characters would put emphasis on Lamarckian possibilities, must be however balanced by anthropological and cultural history studies 'that points to the remarkable persistence and replication of (often tacit) social codes and norms of behaviour', thus maintaining 'some fidelity to the "genotypic" replication of dispositions and rules in social evolution'. (Hodgson, 2003, p. 374).
7. For an analysis of the role of the notion of variety in evolutionary theory, see Hodgson (2003).
8. This principle was originally extended to the entire community, for example to explain the birth of a new collective institution, such as the State in Hobbes. Differently form contractarian approaches, however, deliberation is not a one-off act of delegation.
9. The word is hyphened to put emphasis on the notion of 'liberation' as it happens in the collective context defined by de-liberative processes.
10. 'It is through the camera that I begin to understand the things and the people. That is why I am constantly open to learning and absorbing into the film that which the filming itself reveals, even if that should be in contradiction with what I have written into the script' (Bachmann and Bertolucci, 1973, p. 4).
11. Let us consider again, in this respect, film-making. In Fellini's words, 'Making a film is a metaphor for a type of social utopia: all together doing a thing, directed by one for the good of a cause . . . being involved in realizing a dream. It is the same as a group working on a scientific problem and solving it or involved in a geographical research and discovering a new continent, or the more common ideal of inventing and materializing a social form, working out a philosophical problem or creating a work of art. *The myth of realization* protects by involving you in something bigger than yourself: the aim, the achievement, the goal' (Bachmann and Fellini, 1980, p. 3, emphasis added).

12. 'We are educated to produce, to judge by the achieved, and this causes our unhappiness. If we stopped working for a goal and started working for the sake of work, we might get closer to some form of psychic health. Our functional education claims there is an achievement out there which needs to be reached, but of course there isn't. All there is, is the trip of going out there' (Bachmann and Fellini, 1980, p. 4).
13. Tendencies here refer to events which happen, for instance, as a result of institutional inertia.
14. 'I think it is important to change one's position and focal length from time to time, see things from other angles. Even if this shocks or creates earthquakes in the mind of those who need security, cover, frameworks and roofs . . . He who gives us, through the example of his life or by the expression of his thought and his fantasy, a new view, helping us to pull our concepts out from under the dusty, dim light and out of that small, rational cage of intellect which imprisoned them and kept them from becoming individually significant for us, robbing us, perhaps, briefly, of the consolation of the familiar, of the daily dreariness, and giving back to them a more mysterious meaning of a less predictable sort – I do not think that such a man should be accused of being pessimistic. He is, instead, a realist' (Bachmann and Fellini, 1980–81, p. 7).
15. Compare deliberative democracy with the model of democracy based on preference aggregation. According to Young (2002, p. 20), in the latter 'citizens never need to leave the private realm of their own interests and preferences to interact with others whose preferences differ'. This aggregative model of democracy is said to carry 'a thin and individualistic form of rationality' (ibid.).
16. Harsanyi (1992) distinguishes between 'actual preferences' and 'informed preferences'. Whilst actual preferences are observed in individuals' choices, behaviour and in their verbal statements, informed or true preferences are harder to see as they are inferred from what individuals' preferences '*would* be if they *did* know some pieces of information they actually do not know' (ibid., p. 29).
17. Harsanyi (1992) distinguishes desires from preferences. Whilst preferences provide information about a person's priorities, desires represent a more fundamental notion regarding humans' inner wishes. People may have different preferences on how to accomplish a specific desire, or different preferences regarding the priority of different desires.
18. Acting here is both speculative and physical.
19. For example, it led to: cooperation of academics with Hi8us, a social enterprise located in the UK and dealing with digital media and disadvantaged groups, on the use of digi-essays in academic activity; a 'Workshop on Ethnicity, Creativity and Economic Development', held in Åbo Akademi University, Finland; the design of an innovative curriculum for the Universitas21 Summer School on 'Diversity through Creativity Innovation and Culture' at the University of Birmingham, UK; the 'Mar Del Plata Winter Festival on Economic Development: Opportunities and Choices', held in Argentina; and a planned second Festival on the specific topic of 'Recapturing Space and Time. Creativity and Economic Development' to be held in Italy.
20. This happened namely through the 'Birmingham Workshops on Academic Freedom', as well as through a specific session hosted at the 2008 Annual Conference of the European Network for Industrial Policy (EUNIP).

REFERENCES

Adaman, Fikret and Pat Devine (2001), 'Participatory planning as a deliberative democratic process: a response to Hodgson's critique', *Economy and Society*, **30** (2), 229–39.

Aligheri, Dante (n.d.), 'Inferno', in *The Divine Comedy*, translated with a

commentary by Charles S. Singleton, (1970), Princeton, NJ: Princeton University Press, Canto XXVI, p. 279.
Alchian, A.A. (1950), 'Uncertainty, evolution and economic theory', *Journal of Political Economy*, **58** (June), 211–22.
Arthur, W. Brian (1989), 'Competing technologies, increasing returns, and lock-in by historical events', *Economic Journal*, **99** (1), 116–31.
Bachmann, Gideon and Bernardo Bertolucci (1973), '"Every sexual relationship Is condemned", an interview with Bernardo Bertolucci apropos *Last Tango in Paris*', *Film Quarterly*, **26** (3), 2–9.
Bachmann, Gideon and Federico Fellini (1980–81) '"The cinema seen as a woman . . .": an interview on the day *City of Women* premiered in Rome', *Film Quarterly*, **34** (2), 2–9.
Berlin, Isaiah (1969), *Four Essays on Liberty*, Oxford: Oxford University Press.
Borzaga Carlo and Ermanno C. Tortia (forthcoming), 'The growing social content of cooperative firms: an evolutionary interpretation', in L. Becchetti and C. Borzaga (eds), *The Economics of Social Responsibility*, London: Routledge.
Bruni, Luigino and Robert Sugden (2007), 'The road not taken: how psychology was removed from economics, and how it might be brought back', *Economic Journal*, **117** (516), 146–73.
Chandler, Alfred (1977), *The Visible Hand. The Managerial Revolution in American Business*, Cambridge, MA: Harvard University Press.
Chomsky, Noam (1964), *Current Issues in Linguistic Theory*, 6th edn, 1975, Paris: Mouton.
Chomsky, Noam (1975), 'Prospects for democracy', talk given at MIT, Cambridge, MA, reproduced in C.P. Otero, (ed.) (2003), *Chomsky on Democracy and Education*, London: Routledge Falmer.
Christensen, Jens (2007), 'Thing or mystery? An approach to critical and creative thinking about economic development, democracy and meaning', paper presented at the Ideas Laboratory of the First International Festival on Creativity and Economic Development, Gambettola, Italy, May.
Coase, R.H. (1937), 'The nature of the firm', *Economica*, **IV**, 386–405.
Commission of the European Communities (COM) (2001), 'European governance. A White Paper', COM (2001) 428 Final, Brussels.
Cowling, Keith and Roger Sugden (1998), 'The essence of the modern corporation: markets, strategic decision-making and the theory of the firm', *Manchester School*, **66** (1), 59–86.
Dewey, John ([1910] 1991), *How We Think*, Amherst, NY: Prometheus Books.
Dewey, John (1922), *Human Nature and Conduct: An Introduction to Social Psychology*, New York: Holt.
Dewey, John (1927), *The Public and its Problems*, Denver, CO: Holt; page numbers refer to the reproduction in: J.A. Boydston (1988), *John Dewey. The Later Works Volume 2: 1925–1927*, Carbondale, IL and Edwardsville, LU: Southern Illinois University Press.
Dewey, John ([1934] 2005), *Art as Experience*, New York: Penguin.
Docherty, T. (2008), *The English Question or Academic Freedoms*, Brighton: Sussex Academic Press.
Dosi, Giovanni (1994), 'Firm, boundaries of the', in G.M. Hodgson, W.J. Samuels and M.R. Tool (eds), *The Elgar Companion to Institutional and Evolutionary Economics*, Aldershot, UK and Brookfield, USA: Edward Elgar, pp. 229–37.
Douglas, Mary (1987), *How Institutions Think*, London: Routledge.

Egidi, Massimo (1987), 'Introduzione', to Salvatore S. Rizzello, *L'Economia della Mente*, Rome and Bari, Italy: Laterza.
Elster, Jon (ed.) (1986), *The Multiple Self*, Cambridge: Cambridge University Press and Norwegian University Press.
Harsanyi, John C. (1992), 'Utilities, preferences and substantive goods', WIDER working paper no. 101 (December), United Nations University, Helsinki.
Hayek, Friedrich A. (1945), 'The use of knowledge in society', *American Economic Review*, **35**, 519–30.
Hayek, Friedrich A. (1952), *The Sensory Order. An Inquiry into the Foundations of Theoretical Psychology*, London and New York: Routledge.
Hayek, Friedrich A. (1967), 'The results of human action but not of human design', in Friedrich A. Hayek, *Studies in Philosophy, Politics and Economics*, Chicago, IL: University of Chicago Press.
Hayek, Friedrich A. (1982), *Law, Legislation and Liberty*, London: Routledge.
Hobbes, Thomas ([1651] 1996), *Leviathan*, edited by Richard Tuck, Cambridge: Cambridge University Press.
Hodgson, Geoffrey M. (1993), *Economics and Evolution*, Ann Arbor, MI: University of Michigan Press.
Hodgson, Geoffrey (1998), 'Socialism against markets? A critique of two recent proposals', *Economy and Society*, **27** (4), 450–76.
Hodgson, Geoffrey (2003), 'The mystery of the routine: The Darwinian destiny of an evolutionary theory of economic change', *Revue Economique*, **54** (2), 355–84.
Hodgson, Geoffrey (2005), 'The limits to participatory planning: a reply to Adaman and Devine', *Economy and Society*, **34** (1), 141–53.
Jevons, Stanley H. ([1875] 1905), *Essays on Economics*, London: Macmillan.
Knight, Frank (1921), *Risk, Uncertainty and Profit*, Boston, MA: Houghton Mifflin.
Khurana, R. (2007), *From Higher Aims to Hired Hands. The Social Transformation of American Business Schools and the Unfilled Promise of Management as a Profession*, Princeton, NJ: Princeton University Press.
Langlois, Richard (2003), 'The vanishing hand: the changing dynamics of industrial capitalism', *Industrial and Corporate Change*, **12** (2), 351–85.
Montesquieu, Charles De Secondat ([1748] 1989), *The Spirit of the Laws*, Cambridge: Cambridge University Press.
Myrdal, G. (1957), *Economic Theory and Underdeveloped Regions*, London: Duckworth.
Penrose, Edith Tilton (1952), 'Biological analogies in the theory of the firm', *American Economic Review*, **42** (4), 804–19.
Polanyi, Michael (1967), *The Tacit Dimension*, London: Routledge & Kegan Paul.
Porter, Michael E. (2003), 'The economic performance of regions', *Regional Studies* **37** (6–7), 549–78.
Rizzello, Salvatore (1997), *L'Economia della Mente*, Rome and Bari, Italy: Laterza.
Rostow, W.W. (1959), 'The stages of economic growth', *Economic History Review*, **12** (1), 1–16.
Sacchetti, Silvia and Roger Sugden (2003), 'The governance of networks and economic power: the nature and impact of subcontracting relationships', *Journal of Economic Surveys*, **17**, 669–91.

Sacchetti, Silvia and Roger Sugden (forthcoming). 'The organization of production and its publics: mental proximity, markets and hierarchies', *Review of Social Economy*.
Screpanti, Ernesto (2001), *The Fundamental Institutions of Capitalism*, London: Routledge.
Simon, Herbert A. (1999), 'The many shapes of knowledge', *Revue d'Economie Industrielle*, no. 88, (2nd semester), 23–41.
Spencer, Herbert ([1876] 1969), *Principles of Sociology*, London: Macmillan.
Stabile, Giorgio (2007), *Dante e la Filosofia della Natura. Percezioni, Linguaggi e Cosmologie*, Florence: Sismel.
Sugden, Robert (1991), 'Rational choice: a survey of contributions from economics and philosophy', *Economic Journal*, **101**, 751–85.
Veblen, Thorstein (1898), 'The instinct of workmanship and the irksomeness of labour', *American Journal of Sociology*, **4**(2), 187–201.
Veblen, Thorstein ([1898] 1998), 'Why is economics not an evolutionary science?', *Quarterly Journal of Economics*, July, 373–97, reprinted in *Cambridge Journal of Economics*, **22**, 403–14.
Veblen, Thorstein ([1904] 1996), *The Theory of the Business Enterprise*, New Brunswick, NJ and London: Transaction Publishers.
Williamson, Oliver E. (1985), *The Economic Institutions of Capitalism*, New York: Free Press.
Young, Iris M. (2002), *Inclusion and Democracy*, Oxford: Oxford University Press.

Index